AF800499

Springer Proceedings in Materials

Volume 57

Series Editors

Arindam Ghosh, Department of Physics, Indian Institute of Science, Bengaluru, India

Daniel Chua, Department of Materials Science and Engineering, National University of Singapore, Singapore, Singapore

Flavio Leandro de Souza, Universidade Federal do ABC, São Paulo, Brazil

Oral Cenk Aktas, Institute of Material Science, Christian-Albrechts-Universität zu Kiel, Kiel, Germany

Yafang Han, Beijing Institute of Aeronautical Materials, Beijing, China

Jianghong Gong, School of Materials Science and Engineering, Tsinghua University, Beijing, China

Mohammad Jawaid, Laboratory of Biocomposite Technology, INTROP, Universiti Putra Malaysia, Serdang, Malaysia

Springer Proceedings in Materials publishes the latest research in Materials Science and Engineering presented at high standard academic conferences and scientific meetings. It provides a platform for researchers, professionals and students to present their scientific findings and stay up-to-date with the development in Materials Science and Engineering. The scope is multidisciplinary and ranges from fundamental to applied research, including, but not limited to:

- Structural Materials
- Metallic Materials
- Magnetic, Optical and Electronic Materials
- Ceramics, Glass, Composites, Natural Materials
- Biomaterials
- Nanotechnology
- Characterization and Evaluation of Materials
- Energy Materials
- Materials Processing

To submit a proposal or request further information, please contact one of the following Springer Publishing Editors according to your affiliation location:

European countries: **Mayra Castro** (mayra.castro@springer.com)
India, South Asia and Middle East: **Swati Meherishi** (swati.meherishi@springer.com)
South Korea: **Smith Chae** (smith.chae@springer.com)
Southeast Asia, Australia and New Zealand: **Ramesh Nath Premnath** (ramesh.premnath@springer.com)
The Americas: **Michael Luby** (michael.luby@springer.com)
China and all the other countries or regions, as well as topics in materials chemistry: **Maggie Guo** (maggie.guo@cn.springernature.com)

This book series is indexed in **SCOPUS** and **EI Compendex** database.

Mrutyunjay Suar · Smita Mohanty ·
Namrata Misra · Smruti Ranjan Mohanty
Editors

Advancements in Polymer Technology

Select Proceedings of Polymer Technology Connect 2023

Springer

Editors
Mrutyunjay Suar
Bhubaneswar City Knowledge Innovation Cluster (BCKIC) Foundation
Bhubaneswar, Odisha, India

Smita Mohanty
CIPET: SARP-Laboratory for Advanced Research in Polymeric Materials
Bhubaneswar, Odisha, India

Namrata Misra
Bhubaneswar City Knowledge Innovation Cluster (BCKIC) Foundation
Bhubaneswar, Odisha, India

Smruti Ranjan Mohanty
Bhubaneswar City Knowledge Innovation Cluster (BCKIC) Foundation
Bhubaneswar, Odisha, India

ISSN 2662-3161 ISSN 2662-317X (electronic)
Springer Proceedings in Materials
ISBN 978-981-97-6332-0 ISBN 978-981-97-6333-7 (eBook)
https://doi.org/10.1007/978-981-97-6333-7

© The Editor(s) (if applicable) and The Author(s), under exclusive license to Springer Nature Singapore Pte Ltd. 2024

This work is subject to copyright. All rights are solely and exclusively licensed by the Publisher, whether the whole or part of the material is concerned, specifically the rights of translation, reprinting, reuse of illustrations, recitation, broadcasting, reproduction on microfilms or in any other physical way, and transmission or information storage and retrieval, electronic adaptation, computer software, or by similar or dissimilar methodology now known or hereafter developed.
The use of general descriptive names, registered names, trademarks, service marks, etc. in this publication does not imply, even in the absence of a specific statement, that such names are exempt from the relevant protective laws and regulations and therefore free for general use..
The publisher, the authors and the editors are safe to assume that the advice and information in this book are believed to be true and accurate at the date of publication. Neither the publisher nor the authors or the editors give a warranty, expressed or implied, with respect to the material contained herein or for any errors or omissions that may have been made. The publisher remains neutral with regard to jurisdictional claims in published maps and institutional affiliations.

This Springer imprint is published by the registered company Springer Nature Singapore Pte Ltd.
The registered company address is: 152 Beach Road, #21-01/04 Gateway East, Singapore 189721, Singapore

If disposing of this product, please recycle the paper.

Organization

Advisory Committee

Dr. Ramanuj Narayan, CSIR-Institute of Minerals and Materials Technology, Bhubaneswar, India
Prof. (Dr.) Sanjay K. Nayak, Ravenshaw University, Cuttack, India
Dr. Mrutyunjay Suar, KIIT Deemed to be University, Bhubaneswar, India
Dr. Smita Mohanty, CIPET: SARP-LARPM, Bhubaneswar, India

Convener

Dr. Namrata Misra, Bhubaneswar City Knowledge Innovation Cluster Foundation, Bhubaneswar, India

Coordinators

Dr. Smruti Ranjan Mohanty, Bhubaneswar City Knowledge Innovation Cluster Foundation, Bhubaneswar, India
Dr. Jyoti Agarwal, KIIT-Technology Business Incubation Centre, Bhubaneswar, India

Organizing Committee

Dr. Yatendra S. Chaudhary, CSIR-IMMT, Bhubaneswar, India
Dr. Ratiranjan Nayak, CSIR-IMMT, Bhubaneswar, India

Dr. Sanjib Das, CSIR-IMMT, Bhubaneswar, India
Dr. Akshaya Kumar Palai, CIPET: SARP-LARPM, Bhubaneswar, India
Dr. Saravanakumar B., CIPET: SARP-LARPM, Bhubaneswar, India
Dr. Luna Samanta, Ravenshaw University, Cuttack, India
Dr. Luna Goswami, KIIT School of Chemical Technology, Bhubaneswar, India
Mr. Sudhir Kumar Jha, BCKIC Foundation, Bhubaneswar, India
Mr. Arpan Ghosh, BCKIC Foundation, Bhubaneswar, India
Dr. Jyotsnarani Jena, KIIT-TBI, Bhubaneswar, India
Dr. Aanchal Katoch, BCKIC Foundation, Bhubaneswar, India
Dr. Bhaskar Das, KIIT-TBI, Bhubaneswar, India
Dr. Jayalaxmi Dash, KIIT-TBI, Bhubaneswar, India
Ms. Smrita Pathak, BCKIC Foundation, Bhubaneswar, India
Mr. Malay Mohapatra, KIIT-TBI, Bhubaneswar, India
Ms. Riya Roy, KIIT-TBI, Bhubaneswar, India
Mr. Soumya Ranjan Mahapatra, KIIT School of Biotechnology, Bhubaneswar, India
Ms. Jyotirmayee Dey, KIIT School of Biotechnology, Bhubaneswar, India
Mr. Ray SaiSoubhagya, KIIT-TBI, Bhubaneswar, India
Ms. Shilpa Nayak, KIIT-TBI, Bhubaneswar, India

Sponsors

Preface

It is with great pleasure that we present the proceedings of Polymer Technology Connect (PTC) Conference held on 15th and 16th December. Organized by the Bhubaneswar City Knowledge Innovation Cluster Foundation (BCKIC) and Kalinga Institute of Industrial Technology (KIIT) Deemed to be University, in collaboration with CSIR-IMMT, CIPET Bhubaneswar, and Ravenshaw University, this event served as a dynamic platform for fostering collaboration, exchanging knowledge, and exploring the frontiers of polymer technology.

PTC 2023 brought together industry leaders, academic scholars, and experts from various domains to delve into the diversified areas of polymer science and technology. With thematic areas spanning Polymers in Industrial Applications, Polymers in Biotech and Healthcare, Advanced Polymeric Materials, and Sustainability in the Polymer Domain, the conference provided a comprehensive overview of the latest advancements and emerging trends in the field.

This volume consists of eight research articles selected through a meticulous review process. It is an essential guide for individuals and organizations seeking to navigate the dynamic landscape of polymer technology. It provides in-depth insights into the latest advancements and applications. This volume acts as a bridge, connecting the realms of scientific research and practical industrial applications. It explores cutting-edge polymeric materials, their impact on diverse industries, and their role in shaping a more sustainable future. Readers will find valuable insights into smart polymers, nanocomposites, biodegradable materials, and much more, providing a glimpse into the future of this ever-evolving field.

We extend our sincere gratitude to Springer, our esteemed publication partners, for their unwavering support in bringing this volume to fruition. Additionally, we express heartfelt appreciation to our sponsors—DST-SERB, KINGFA, ACS, Eppendorf, Stratasys, RSC, and Inkarp—for their invaluable contributions, which have played a pivotal role in enhancing the quality and impact of the conference. Special thanks to the keynote presenters, and eminent experts whose insights enriched the conference sessions. We are immensely grateful to the authors and reviewers for their dedication and cooperation, without which this volume would not have been possible.

Bhubaneswar, India

Mrutyunjay Suar
Smita Mohanty
Namrata Misra
Smruti Ranjan Mohanty

Contents

About the Editors

Dr. Mrutyunjay Suar Chairman of the Bhubaneswar City Knowledge Innovation Cluster (BCKIC) Foundation, is also the Director General of R&D and Innovation at KIIT University and CEO at KIIT-TBI. Under his leadership, KIIT-TBI has nurtured 400+ start-ups, creating 6000+ jobs. He holds a master's in marine biotechnology, a Ph.D. in molecular microbiology, and has postdoc experience at ETH Zurich. Dr. Suar founded two companies, holds six patents, and has over 150 publications in infection biology and nano-biomaterials.

Dr. Smita Mohanty Director and Head of CIPET: SARP-LARPM, is a leading academician with over 15 years of research and teaching experience. Under her visionary leadership, LARPM has advanced in E-waste Recycling and Biopolymer development. She has published over 100 research articles in international journals and holds five Indian patents, showcasing her innovative contributions in polymer sector. Dr. Mohanty has also completed several industrial-sponsored projects.

Dr. Namrata Misra CEO of BCKIC Foundation, is also an Advisor of KIIT TBI, and Associate Professor at KIIT School of Biotechnology. Besides her leadership role in innovation and start-up ecosystem, she is active in research and her work includes around 70 international publications, the development of computational tools, and editorial roles in scientific books. Dr. Misra is an awardee of the "Odisha Young Scientists Award 2019" and the "Women Excellence Award 2022".

Dr. Smruti Ranjan Mohanty is the Lead of the Technology Partnership at BCKIC Foundation. Before that, he was working as the Assistant Manager at SRF Ltd. He has a Ph.D. in Polymer Technology and his core research expertise is focused on new thermoplastics and polyurethane products. His accolades include the prestigious 10th National Award for Technology Innovation and the Excellent New Staff of the Year 2021 at Kingfa.

Development of Polydimethylsiloxane and Polyurethane (PU/PDMS) Based Coating for Foul-Release Applications

Smruti R. Mohanty, Priyanka Pandey, Smita Mohanty, and S. K. Nayak

Abstract The current research work deals with the development of environment-friendly coatings with fouling release ability based on polydimethylsiloxane (PDMS) and polyurethane (PU) for marine applications. PU was synthesized using isophorone diisocyanate (IPDI) and polyethyleneglycol (PEG) and blended with PDMS. FTIR spectroscopy confirms the synthesis of PU and the successful interaction of PDMS components with the PU domain. The properties of PU/PDMS based films were evaluated in terms of mechanical strength, solubility, chemical resistance, water absorption, thermal stability, glass transition temperature (Tg), and morphology. The NCO:OH ratio of 1.4:1 and PU:PDMS ratio of 85:15 was found to be the optimum composition. GPTMS-treated nanosilica was also incorporated at a loading of 2 wt% within the PU/PDMS system. Subsequently, the optimized formulation of PU/PDMS was coated on metal substrates and exposed to marine water containing algal colonies. The modification of the PU/PDMS blend with nanosilica was also investigated to evaluate the quality of the coating.

Graphical Abstract

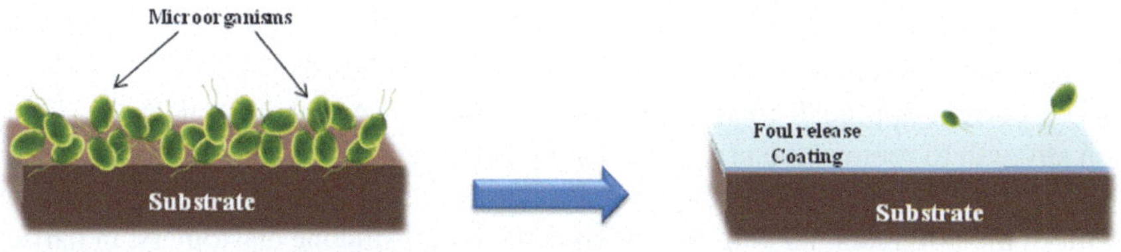

Keywords Marine biofouling · Foul release coating · Non-biocidal · PDMS · PU

S. R. Mohanty (✉) · S. Mohanty · S. K. Nayak
CIPET: SARP-Laboratory for Advanced Research in Polymeric Materials (LARPM), B - 25, CNI Complex, Bhubaneswar, Odisha 751024, India
e-mail: 108smruti@gmail.com

P. Pandey
National Institute of Science Education and Research, Bhubaneswar, P.O. Jatni, Khurda, Odisha 752050, India

© The Author(s), under exclusive license to Springer Nature Singapore Pte Ltd. 2024

M. Suar et al. (eds.), *Advancements in Polymer Technology*, Springer Proceedings in Materials 57, https://doi.org/10.1007/978-981-97-6333-7_1

1 Introduction

Marine structures such as ships and platforms are under constant attack from the marine environment such as saltwater and biological attack by a wide variety of microorganisms that lead to biofouling [1]. The control of marine biofouling on submerged man-made surfaces has become a prominent concern in recent years. In the sea, biofouling is caused by micro-organisms of above 3,500 species, including bacteria, viruses, protozoa, fungi, and algae [2]. A routine sequence of occurrences was followed for the development of biofouling in aquatic environments: (1) blooming of conditioning film, (2) settling of microorganisms, (3) algal and larvae colonization, and (4) extension of large organisms [3]. The conditioning film was eventually developed with the adhesion of subatomic particles and organic molecules such as polysaccharides and proteins on the surface of the substrates underwater.

Coatings based on wax, pitch, and tar have a long history and have been primarily used as an important option for controlling macro fouling on marine vessels for over 2000 years [4]. For the last 200 years, the utilization of copper in antifouling coatings to cease the growth of microorganisms has been widely used but the corrosion caused by it stimulates the development of polymer coating. Tributyl-tin containing antifouling coatings were initially patented in the mid-1970s [5]. Due to the discharge of arsenical substances, toxicity, and consequent ecological impacts, their domination was ended as a result of the ban on the use of tin-based coatings on 1 January 2008 [6].

In the worldwide coatings market for marine, biocide-based antifouling (AF) and foul release coatings (FRC) were pre-eminent. The growth of foul release coatings increased significantly to 10% by volume for marketable shipping after the adoption of the IMO convention in 2003. About 70 million tons of copper-based chemicals and 20 million liters of solvent per annum were estimated to be saved with the total replacement by FRC. The FRC is advantageous due to a 20–30% higher volume of solid content with the help of emission reduction in shipyards [7].

The current approach to foul release was the coatings that weaken the adhesion strength to remove the organisms easily with the hydrodynamic force of water. The most likely pretender for broad-scale non-fouling materials is amphiphilic or zwitterionic polymers made from polysiloxane based silicone elastomers, in particular PDMS. It has a good elasticity and low surface energy, making the organism isolated from the coating surface under shear stress initiated by hydrodynamic drag [8]. However, since PDMS has a weak mechanical strength and modulus, it cannot be used as a sole material for coating beneath the ships [9]. Polyurethanes paired with polysiloxane present notable advantages such as enhanced mechanical strength, elasticity, resistance to adhesion, and abrasion in foul-release coatings. Nevertheless, achieving uniform physical blends of PDMS and polyurethanes poses a challenge due to the inherent incompatibility of these resins, leading to a marked tendency for phase separation [10]. Nano silica particles are broadly used in the coatings sector due to their high resistance to fouling in marine environments. In order to raise the application performance, their surfaces usually need to be modified with silane to increase the hydrophobicity of the film [11].

Baier proposed that the superficial adhesion of protein fouling is exhibited at a surface tension of around 20 mN/m, [12] which is the surface tension of pure PDMS [13]. Holm et al. reported that the ideal substrate for one fouling type may not be the same for another [14]. Several similar reports [15–17] were available pertaining to the use of PU/PDMS as a suitable material for marine applications. The green alga Ulva linzawas was shown to have a lower affinity to hydrophobic substrates while diatoms showed strong adhesion to hydrophobic surfaces. Also, there is a difference in protein compositions and sequence between the barnacles B. eburneus and B. crenatus the mussel M. edulis. Thus, solving the problem with adhesion for one species may not be the ideal solution for another.

In the current study, zwitterionic polymer using Polyurethane (PU) and Polydimethylsiloxane (PDMS) were synthesized for foul release applications in the marine environment. Functionalized nanosilica was also incorporated to enhance the strength and foul release properties of the coating. An in-depth study on the physicomechanical, thermal, morphology, and foul release properties of the blends has been investigated.

2 Materials

Polyethylene glycol (PEG) (Mol wt. 1000 g/mol^{-1}) was received from M/s HIMEDIA Laboratories, Mumbai, India. Isophorone diisocyanate (IPDI) (Mol wt. 222.29 g/mol^{-1}) and Dibutyltin dilaurate (DBTDL) (Mol wt. 631.57 g/mol^{-1}) were purchased from M/s TCI Chemicals (India) Pvt. Ltd., Chennai, India. Aminopropyl terminated Polydimethylsiloxane (APT-PDMS) (Mn = 3000) was acquired from M/s Gelest, Inc., PA, USA. 1,4-Butanediol (BD) (Mol wt. 90.12 g/mol^{-1}), (3-Glycidyloxypropyl) trimethoxysilane (GPTMS) (>= 98%), and Silica nanopowder were obtained from M/s Sigma-Aldrich, Missouri, USA. Acetone was provided by M/s SRL Pvt. Ltd, Mumbai, India. All other solvents of analytical reagent grade were used without any modification.

3 Experimental

3.1 Synthesis of Polyurethane (PU)

Polyurethane (PU) was synthesized using Poly-ethylene glycol (PEG) and Isophorone diisocyanate (IPDI) at NCO: OH ratio of 1.4:1. In order to synthesize PU, PEG (dehumidified in a vacuum oven at a temp of 60 °C for 30 min) and IPDI each dissolved in 5 ml acetone was stirred in two 250 ml round-bottomed flasks equipped with a magnetic stirrer and stir bar separately at 500 rpm for 1 h at room temperature and controlled atmospheric conditions. Subsequently, 0.01 wt% of DBTDL was added to IPDI to increase the reaction rate. Both solutions were mixed in a flask and stirred continuously for 4 h. 0.02 wt % of 1, 4-Butanediol was added to increase the cross-linking. The viscosity of the solution was controlled at an optimized temperature of 80 °C till the solvent get evaporated completely. Subsequently, the solution was poured on a glass plate and cured at room temperature for 4–5 days.

3.1.1 Synthesis of PU/PDMS Blends and its Nanocomposite

A blend of PU and PDMS was prepared at wt. ratio of 85: 15 based on previous literature [18]. PDMS was initially added to the IPDI solution in acetone and stirred for 1 h in a 250 ml round-bottomed flask before the addition of the PEG solution. Then the catalyst DBTDL and 1,4-Butanediol were added and the process was completed similarly as described above. Further, the PU/PDMS nanocomposites were also prepared using nanosilica (NS) and modified nanosilica (MNS) as reinforcing agents. The compositions prepared are presented in Table 1.

4 Characterization

4.1 Structural Analysis

FTIR spectra were recorded on Nicolet 6700, USA with a scan range of 400–4000 cm^{-1} and 0.1% transmission accuracy.

Table 1 Composition of different PU/PDMS blends

Blend type	PU (wt%)	PDMS (wt%)	NS (wt%)	MNS (wt%)
PU	100	0	0	0
PU/PDMS	85	15	0	0
PU/PDMS/NS	85	15	2	0
PU/PDMS/MNS	85	15	0	2

Note PU = Polyurethane, PDMS = Aminopropyl terminated Polydimethylsiloxane, NS = Nanosilica, MNS = Modified Nanosilica

4.2 Mechanical Properties

The mechanical characteristics were evaluated using Universal Testing Machine (UTM), 3382, Instron, UK, in accordance with ASTM D 2370 at a crosshead speed of 0.5 mm min^{-1} and gauge length of 100 mm.

4.3 Water Absorption

The water absorption test was performed in compliance with ASTM D 570 and was calculated as per Eq. 1. Test specimens having 50 mm diameter and 2 mm thickness were immersed in distilled water at room temperature. The specimen shall be removed from water and weighed at an interval of every 2 h for 6 h.

$$\%\ \text{Water Absorption} = \frac{Wet\ weight(w_2) - Dry\ weight(w_1)}{Dry\ weight(w_1)} \times 100 \quad (1)$$

4.4 Chemical Resistance

The chemical resistance of the films against Hydrochloric acid (HCl), Hydrofluoric acid (HF), Potassium hydroxide (KOH), and Ammonium hydroxide (NH_4OH) were studied by dipping a piece of each sample in these acidic and basic solutions at room temperature.

4.5 Swelling Behavior

The swelling behavior was carried out with various solvents such as Tetrahydrofuran (THF), Dimethylformamide (DMF), Dichloromethane (DCM), and Methyl Ethyl Ketone (MEK).

4.6 Crosslink Density

The crosslink density of PUs was obtained according to the Flory–Rehner Equation as shown in Eq. 2:-

$$v = \frac{\ln(1 - v_r) + v_r + \chi v_r^2}{v_s\left[v_r^{1/3} - 0.5v_r\right]} \quad (2)$$

where, v is the crosslink density, v_r is the volume fraction of coating material in the swollen mass, v_s is the mole volume of used solvent and χ is the Flory solvent-polymer interaction term.

4.7 Differential Scanning Calorimetry (DSC)

DSC was performed on DSC Q 20, TA Instruments, USA in a temperature range of −50 to −120 °C at a heating rate of 10 °C/min under an inert nitrogen atmosphere to record the glass transition temperature.

4.8 Thermal Degradation by Thermogravimetric Analysis (TGA)

5 gm of samples were investigated on TGA Q 50, TA Instruments, USA at a temperature range from 100 to 600 °C for the study of thermal degradation behavior.

4.9 Surface Topography

Surface topography and heterogeneity were studied on AFM, XE-100, PARK-South Korea with a maximum scan size of 50 μm at an ambient temperature of 25 ºC.

4.10 Algae Adsorption

The formulations were coated on steel plates and immersed in seawater collected from the Bay of Bengal, Paradeep, India, which contains fouling alga *Ulva spores* and several biological organisms such as *E. coli, P. aeruginosa, S. epidermidis, B. subtilis C. albicans, S. aureus*, etc. for 10 days.

5 Results and Discussion

5.1 Fourier Transform Infrared Spectroscopy (FTIR)

The FTIR pattern of IPDI exhibited a characteristic peak at 2955 cm^{-1} indicating the presence of isocyanate group (–NCO). Further, the polyurethane prepared at NCO:OH ratio of 1.4:1 exhibited additional peaks at 3369 cm^{-1} (N–H stretching), 2867 cm^{-1} (–CH_2 and –CH_3 stretching vibration), 1713 cm^{-1} (H-bonded –C = O of urethane linkage) and 1554 cm^{-1} (C–NH-vibration), indicated the successful synthesis of polyurethane via formation of urethane linkages. H-bonded –C = O urethane linkages, indicate the formation of H-bond between NH (proton donor) and C = O (proton acceptor). This is because the urethane groups possess charged H, N, and O atoms that act as dipoles and hence attract oppositely charged atoms [19] (Fig. 1).

Further, it was noticed that the intensity of peak at 2955 cm^{-1} (–NCO), was reduced in the case of PU (1.4:1) thereby representing the utilization of few isocyanate groups in the formation of polyurethane and some amount of NCO is unreacted. Therefore, it may be confirmed that the polyurethane prepared at an NCO: OH ratio of 1.4:1 is isocyanate-terminated with the appropriately formed urethane linkages. Hence this composition was considered for further evaluation as PDMS is expected to interact with the polyurethane via its –NCO group.

However, the peak corresponding to –NCO was absent in the case of PU/PDMS (85:15), thus, indicating that the entire –NCO group present in IPDI has been

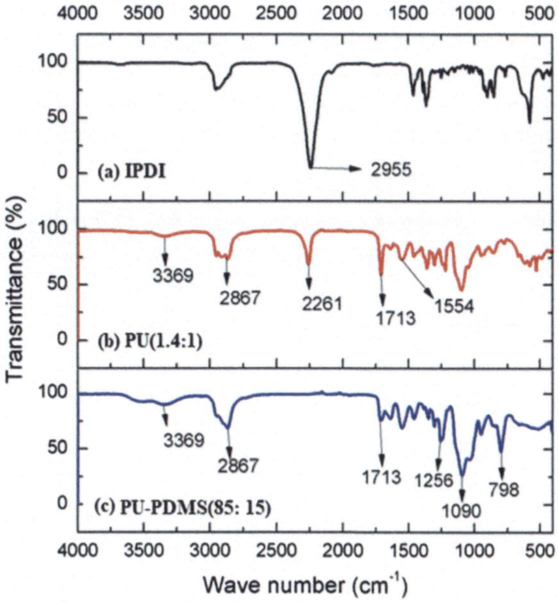

Fig. 1 FTIR spectrum of **a** IPDI **b** PU(1.4:1) **c** PU-PDMS(85:15)

completely utilized. The band corresponding to 3369 cm^{-1} confirms the formation of amide structure between PDMS and the PU matrix. The bands at 798 cm^{-1} indicate the presence of a Si–C bond, peak at 1256 cm^{-1} revealed the presence of symmetrical CH_3 groups present in PDMS. Similarly, peaks at around 1090 cm^{-1} represent overlapped bands of Si–O–Si and C–O–C groups, respectively, thereby confirming the successful interaction of the PDMS component with the polyurethane domain [20].

5.2 Mechanical Characterization

As displayed in Table 2, the tensile strength and tensile modulus of PU were found to be 0.287 and 1.14 MPa respectively. Incorporation of PDMS leads to a reduction in all these properties that can be accredited to the enhancement of soft segment content that in turn leads to lower mechanical properties. Further, incorporation of NS also does not impart any noticeable improvement suggesting the incompatibility of NS within the PU/PDMS system. Moreover, the incorporation of GPTMS-modified nanosilica (MNS) showed a considerable improvement of all the parameters as compared with that of PU/PDMS and PU/PDMS/NS indicating the improved compatibility due to efficient nanosilica tethering to the matrix polymer as well as hydrogen bonding between NS particles and the polymer after modification with GPTMS. Moreover, it is noticed that the tensile strength and tensile modulus of PU/PDMS/MNS were lower as compared with that of PU to the tune of 53% and 73%. This was ascribed to the restricted mobility of the molecular chains due to the tethering of NS particles. However, the mechanical finding obtained does not appreciably affect the use of PU/PDMS/MNS for foul-release coating application, since the required tensile modulus for foul-release coating is <1 MPa. However, the elongation at break of PU/PDMS/MNS increased to the tune of 63% as compared with that of PU. This reveals the higher toughness of PU/PDMS/MNS as compared to other systems [21].

Table 2 Tensile parameters for PU, PU/PDMS, PU/PDMS/NS and PU/PDMS/MNS

Sample	Tensile strength (MPa)	Elongation at break (%)	Tensile modulus (MPa)
PU	0.287 ± 0.1	868 ± 3.5	1.14 ± 0.6
PU-PDMS	0.050 ± 0.01	222 ± 5.8	0.45 ± 0.02
PU/PDMS/NS	0.065 ± 0.01	63 ± 7.6	0.27 ± 0.02
PU/PDMS/MNS	0.075 ± 0.02	931 ± 2.2	0.53 ± 0.1

5.3 Water Absorption

The water absorption of PU, PU/PDMS, PU/PDMS/NS, and PU/PDMS/MNS systems were represented in Fig. 2. As PDMS was introduced into PU, it was found that the water absorption increased to the tune of 35.8% in 300 min which is possibly due to higher small soluble molecules produced as aminopropyl terminated PDMS which probably have some termination effects. The absorption in the NS-based film was quite high as compared to all other formulations due to the hydrophilic nature of nanosilica [22]. The PDMS soft segment tends to migrate towards the surface layer due to its significantly lower affinity for water compared to the PU chain in the process of film formation [23]. Further, the incorporation of modified nanosilica decreases the water absorption of the film to the tune of 12.2% due to the tough hydrophobic nature of the –Si-OCH_3 bond present in the GPTMS moiety.

Based on the above observations, out of all the compositions, PU, PU/PDMS, and PU/PDMS/MNS were considered for further investigation.

5.3.1 Chemical Resistance

Chemical resistance of PU, PU/PDMS, and PU/PDMS/MNS was studied in terms of loss in weight %, and the results are outlined in Table 3.

In the acid media, all the samples exhibited good stability in their physical appearance, weight, and color. However, in alkaline media in which KOH and NH_4OH, PU exhibited yellowing due to the presence of labile urethane linkages of PU causing it to undergo base hydrolysis that in turn leads to the discoloration and formation

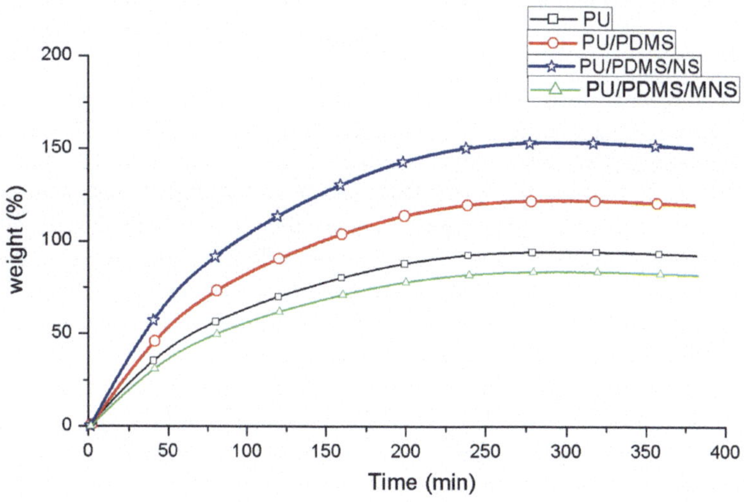

Fig. 2 Water absorption graph of PU, PU/PDMS, PU/PDMS/NS, PU/PDMS/MNS

Table 3 Behaviour of PU, PU/PDMS, and PU/PDMS/MNS upon exposure to various chemicals

Chemical	PU	PU/PDMS	PU/PDMS/MNS
HCl	No change	No change	No change
HF	No change	No change	No change
KOH	Color change with Yellowing	No color change/swelling	No change
NH_4OH	Color change with Yellowing	No color change/swelling	No change

of cracks. Conversely, the PU/PDMS samples in alkaline media showed no color change, however, the swelling was noticed that may be ascribed to the interaction between the PEG and OH group of the solvent media. Furthermore, no swelling in the PU/PDMS/MNS samples thus suggesting that modified nanosilica has effectively enhanced the crosslinked network of polymer systems, owing to the H-bonds forms between silica and polyurethane network. This also indicates that the silica particles act as the bridge between constituents of polymer systems that in turn prevent the migration of solvent [24].

5.4 *Swelling Behavior and Crosslink Density*

The swelling assessment has been carried out for PU, PU/PDMS, and blends and their nanocomposites in a variety of solvents having dissimilar solubility parameters value (δ) ranging from 7.3 to 16.5 $(cal/cm^3)^{1/2}$. The highest peak analogous to the solubility parameter of tetrahydrofuran (THF) (δs) was observed, as there was the highest swelling merely in this solvent. The polyurethane systems do not diffuse in a non-reactive solvent and only reveal their cross-linked nature. The order of swelling capacity of polyurethane in various solvents was found to be: DMF < MEK < DCM < THF.

In a non-reactive solvent, the degree of swelling regulates the degree of cross-linking and consequently the molecular weight between cross-links (M_c). Elevated the M_c, inferior will be the crosslink density. Further, the values of the equilibrium degree of swelling of the PU, PU/PDMS, and PU/PDMS/MNS evaluated in THF are summarized in Table 4. The equilibrium swelling degree of PU increased from 0.783 to 1.889 (in THF) with the incorporation of PDMS, since it forms the soft segment within the PU. The crosslink density of the synthesized PUs was calculated using the results obtained from equilibrium swelling in THF according to the Flory–Rehner Equation. It was observed that the crosslink density of the PU was marginally increased with the incorporation of PDMS content that may be ascribed to the crosslinking within the PU and PDMS components. Furthermore, the incorporation of MNS within the PU/PDMS systems exhibited a crosslink density of 25.65×10^4 which was higher to the tune of 5.5% as compared with that of PU/PDMS which further ascribed that the incorporation of nanosilica enhances the compatibility and

Table 4 The equilibrium swelling degree (q_e) and crosslink density (ν) of PU, PU/PDMS and PU/PDMS/MNS

Sample	q_e	$\nu \times 10^4$/mol cm^{-3}
PU	0.783	25.87
PU/PDMS	1.889	24.31
PU/PDMS/MNS	0.942	25.65

cross-linking between PU and PDMS, thereby preventing the migration of solvent as well as swelling of the sample [25].

5.5 Differential Scanning Calorimetry (DSC)

The glass transition temperature (T_g) for all the systems was determined using DSC analysis and the resulting thermogram is presented in Fig. 3.

Polyurethane exhibited two glass transition temperatures at −17 and 42 °C corresponding to the soft segment (T_{gsoft}) and hard segment (T_{ghard}) respectively which revealed the phase segregation between the hard and soft segments. Further, the incorporation of PDMS results in the shift of T_{gsoft} to the higher temperature of 19 °C, which reveals the incompatibility of PDMS with the soft segment of polyurethane, since PDMS contributes to the soft segment of the polyurethane system. Moreover,

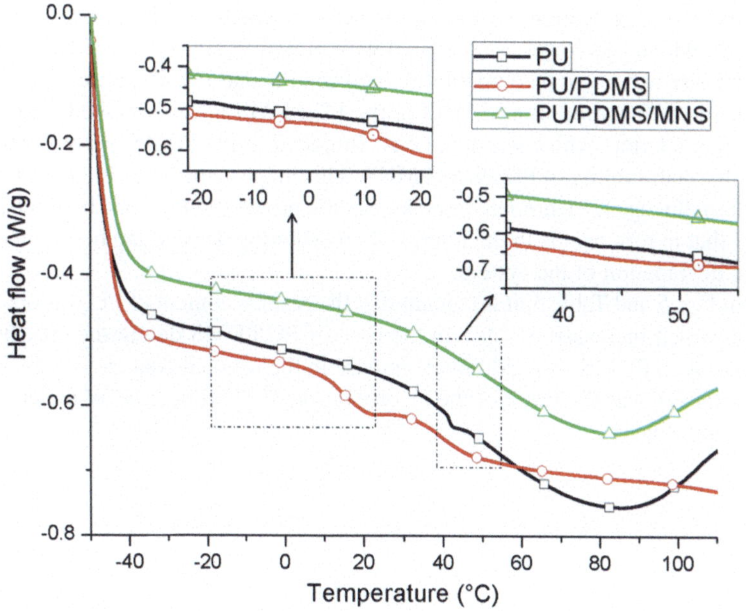

Fig. 3 DSC thermogram of PU, PU/PDMS and PU/PDMS/MNS

no change in the T_{ghard} revealed the unaltered morphology of urethane linkages within the hard segment. Further, the incorporation of MNS within the PU/PDMS system showed the absence of T_{gsoft} transition that may be ascribed to the shift of T_{gsoft} to the temperature below −50 °C. This is attributed to the improved compatibility between PDMS and the soft segment of PU owing to the formation of Si–O–Si linkages [26].

Further, in the case of PU/PDMS/MNS, the presence of terminal ethoxy units in GPTMS prepolymer improves miscibility between reactants and therefore, results in lower T_{ghard} values to the temperature of 37 °C and consequently, a lower degree of microphase separation. Further, in all the cases, no melting transition was observed below 100 °C.

5.6 Thermogravimetric Analysis (TGA) of PU, PU/PDMS and Blends and its Nanocomposites

The thermal degradation behavior of PU, PU/PDMS, and blends and its nanocomposites evaluated using TGA are presented in Fig. 4 and Table 5. PU showed a two-step degradation corresponding to the hard segment at 365 °C and the soft segment at 411 °C. Two distinct peaks represent the phase-segregated morphology of the polyurethane system [27]. Further, it was noticed that 50% weight loss occurred near 362 °C. However, in the case of the PU/PDMS blend, peaks corresponding to the soft and hard segments were not appropriately distinct, representing the lower degree of micro-phase segregation. Further, the onset of degradation (T_{donset}) was found to be 326 °C which was lower as compared with that of PU (T_{donset} = 364 °C), which might be ascribed to the incompatibility between PU and PDMS phases.

Further, the incorporation of MNS in the PU/PDMS blend exhibited a T_{donset} of 333 °C degradation, which was higher as compared to PU/PDMS, representing the higher thermal stability of PU/PDMS/MNS. This may be attributed to the formation of Si–O–Si linkage [28] through the condensation process and H-bond between MNS and PU that in turn results in the improved crosslinking density and hence hindered thermal degradation of the system.

From Fig. 5 and Table 5, it is evident that the residue content of PU was obtained as 0.3% which increased to 4.9% in the case of PU/PDMS due to the presence of the Si group in PDMS. An additional increase in the residual content to 5.2% in the nanocomposite was the result of the incorporation of MNS because of the presence of silica and silane.

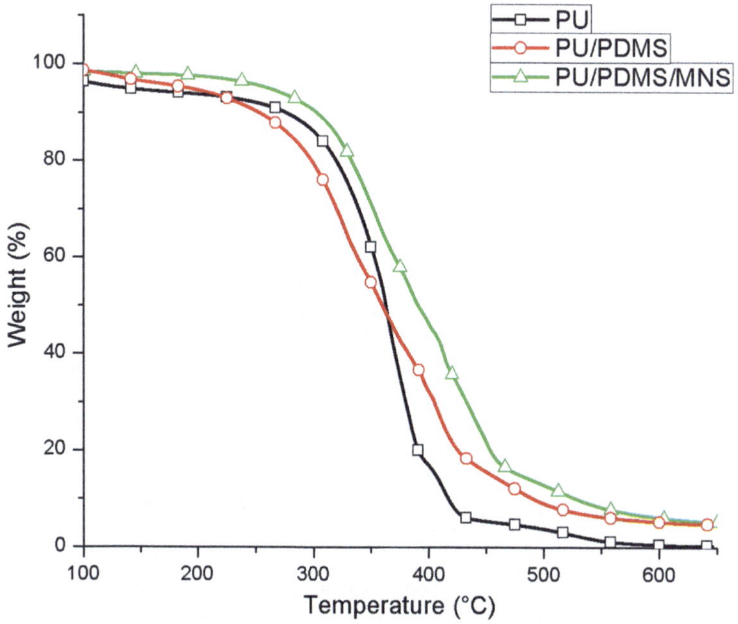

Fig. 4 TGA thermogram of PU, PU/PDMS and PU/PDMS/MNS

Table 5 Thermal degradation temperature of Polyurethane systems

Sample	T_{donset}(°C)	T_{dend}(°C)	$T_{d50\%}$	Residue (%)
PU	364	412	362	0.3
PU/PDMS	326	407	359	4.9
PU/PDMS/MNS	333	385	364	5.2

5.7 *Atomic-Force Microscopy (AFM) Analysis of PU/PDMS/MNS*

AFM analysis was conducted to investigate the integration of modified nanosilica and measure the influence of PDMS soft and hard segments on the superficial topography and liberation of heterogeneity of the film. The findings are demonstrated in Fig. 6. It is known that the light brown zones represent the soft segment, while the darker zones represent the hard segment in polyurethane based on prior studies; Further, the bright particulate formation indicates the presence of nanosilica on the surface.

Insertions of the three-dimensional extensive resolution images clearly illustrate that the surface structure changes and the roughness diminishes with the incorporation of modified nanosilica and occupancy of PDMS soft segments. Based on surface conformation, these results are in agreement with previous reports [29].

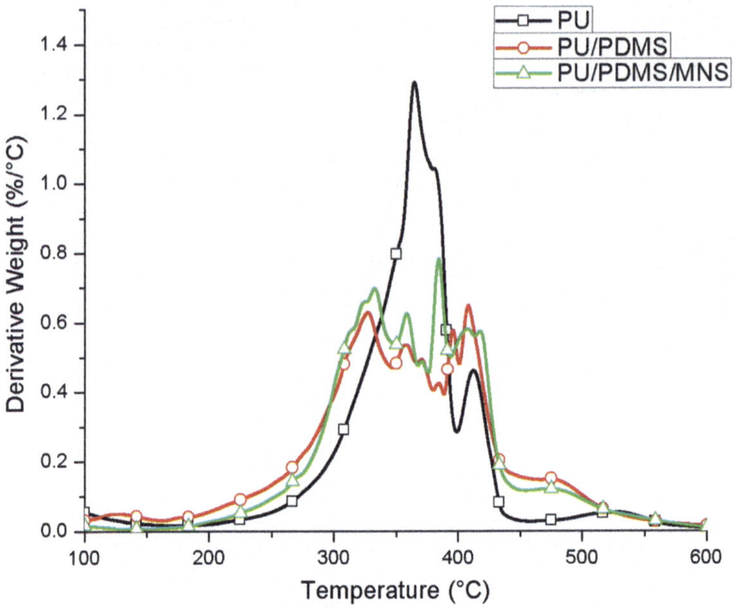

Fig. 5 DTG thermogram of PU, PU/PDMS, PU/PDMS/NS and PU/PDMS/MNS

5.8 Algae Adsorption and Release Analysis

The PU/PDMS and PU/PDMS/MNS formulations were coated on a steel alloy plate. It was noticed that the adhesive property of the film increases with the incorporation of MNS. The coated substrates were exposed to marine water containing algal colonies such as *Ulva spores* and several biological organisms such as *E. coli, P. aeruginosa, S. epidermidis, B. subtilis C. albicans, S. aureus*, etc. [30] for 10 days which revealed lower accumulations of algal colonies in case of PU/PDMS/MNS as compared with that of the PU/PDMS system (Fig. 7).

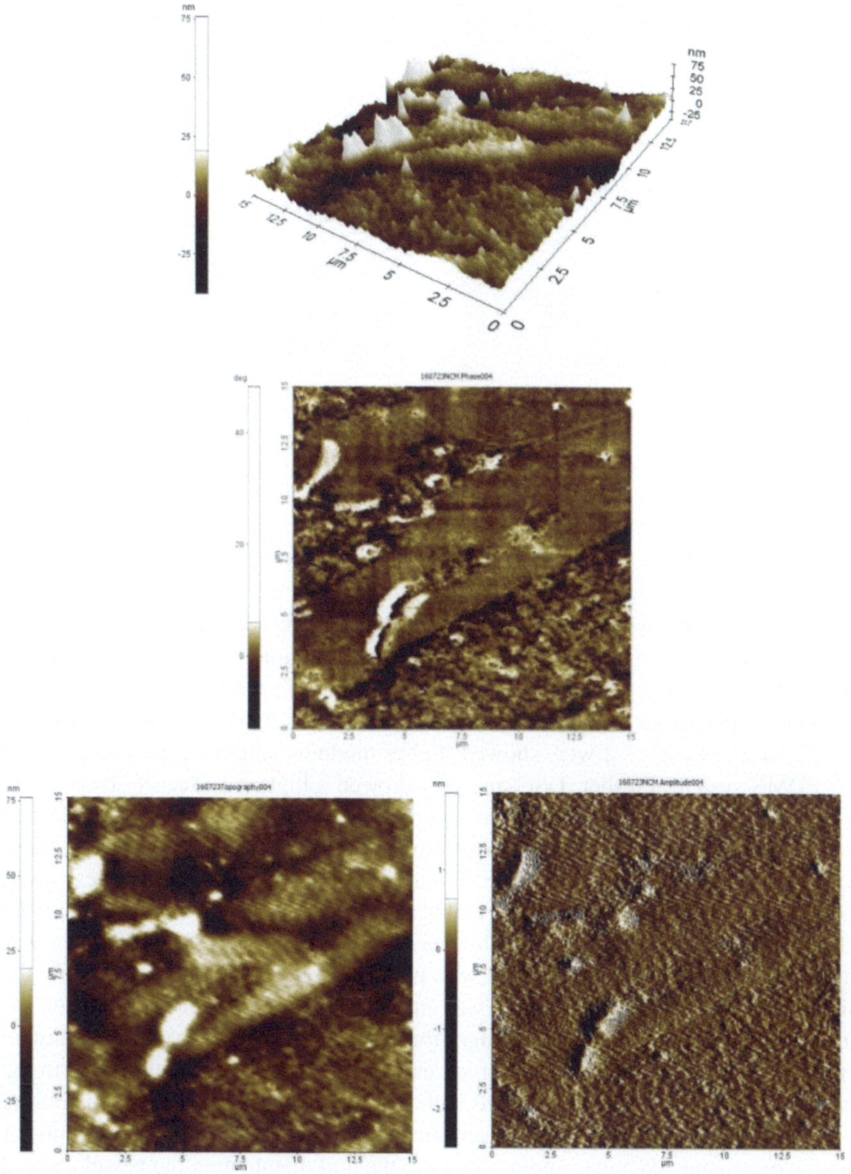

Fig. 6 Topography images of PU/PDMS/MNS by AFM

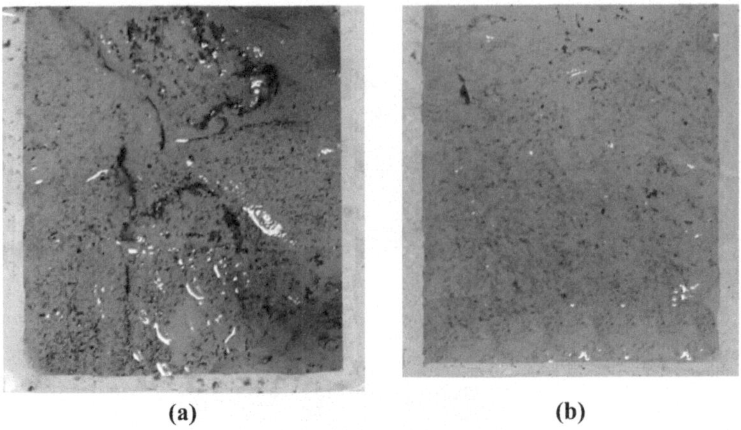

Fig. 7 Algal adsorption on **a** PU/PDMS and **b** PU/PDMS/MNS coating

6 Conclusion

A series of PDMS blended PU was synthesized. The NCO: OH ratio of 1.4:1 and PU: PDMS ratio of 85:15 was found to be the optimum composition. The incorporation of untreated nanosilica did not reveal any noticeable improvement in the mechanical properties of PU, however, the addition of GPTMS-modified nanosilica (MNS) at a loading of 2 wt% showed higher modulus and elongation at break of PU/PDMS system. Further, DSC analysis showed a higher degree of crosslinking within the PU/PDMS/MNS system as compared with that of PU and PU/PDMS and a relatively lower degree of phase separation. However, a certain amount of phase separation was evident from the AFM analysis. Water absorption, chemical resistance, and swelling properties also revealed a higher degree of crosslinking in the PU/PDMS/MNS system. Furthermore, the thermal stability of the PU/PDMS/MNS was also found to be higher as compared with that of PU and PU/PDMS confirming the higher degree of crosslinking which was attributed to the compatibility between PU and PDMS, thereby preventing the migration of solvent as well as swelling of sample. Further, the accumulation of algal colonies was found to be relatively lower in the case of PU/PDMS/MNS as compared with that of PU/PDMS systems. Anti-fouling coatings are applied to the submerged areas of ships, to prevent the adhering of marine organisms like algae and mollusks to the hull maintaining the vessel's speed and reducing fuel consumption. Foul-release coatings offer an additional advantage by safeguarding the integrity of metal surfaces, unlike traditional antifouling methods, which can compromise material strength. This distinction makes foul-release coatings particularly well-suited for protecting propellers, running gear, and other underwater metal components.

Acknowledgements This study was supported financially by DST-TSD, New Delhi, Govt. of India which is gratefully acknowledged.

References

1. Callow JA, Callow ME (2011) Trends in the development of environmentally friendly fouling-resistant marine coatings. Nat Commun 2(1):244. https://doi.org/10.1038/ncomms1251
2. Lakshmi VV, Akondi KB (2015) Control of biofouling in marine environment—past, present and future. Int J Sci Eng Res 6(2):283–286
3. Zhang Y, Qi Y, Zhang Z (2016) Synthesis of PPG-TDI-BDO polyurethane and the influence of hard segment content on its structure and antifouling properties. Prog Org Coat 97:115–121. https://doi.org/10.1016/j.porgcoat.2016.04.002
4. Yebra DM, Kiil S, Dam-Johansen K (2004) Antifouling technology—past, present and future steps towards efficient and environmentally friendly antifouling coatings. Prog Org Coat 50(2):75–104. https://doi.org/10.1016/j.porgcoat.2003.06.001
5. Wells S, Sytsma M (2009) A review of the use of coatings to mitigate biofouling in freshwater. Portland State University, Portland, OR. Accessed 22 Feb 2018
6. Coast A (2013) International convention 4(3)
7. Lejars M, Margaillan A, Bressy C (2012) Fouling release coatings: a nontoxic alternative to biocidal antifouling coatings. Chem Rev 112(8):4347–4390. https://doi.org/10.1021/cr200350v
8. Popa PJ, Li Y, Chen H (2015) Siloxane-urethane foul release coatings. U.S. Patent Application No. 14/795,088
9. Pieper RJ, Ekin A, Webster DC, Cassé F, Callow JA, Callow ME (2007) Combinatorial approach to study the effect of acrylic polyol composition on the properties of crosslinked siloxane-polyurethane fouling-release coatings. J Coat Technol Res 4:453–461
10. Shabel BS, Topolosky GP (2001) (19) United States correspondence address, vol 1, no 09, pp 1999–2001
11. Li J, Zhao Y, Hu J, Shu L, Shi X (2012) Anti-icing performance of a superhydrophobic PDMS/modified nano-silica hybrid coating for insulators. J Adhes Sci Technol 26(4–5), 665–679. https://doi.org/10.1163/016942411X574826
12. Baier RE (2014) Correlations of materials surface properties with biological responses. J Surf Eng Mater Adv Technol 5(01):42. https://doi.org/10.4236/jsemat.2015.51005
13. Owen MJ (1988) Surface tension of polytrifluoropropylmethylsiloxane. J Appl Polym Sci 35(4):895–901. https://doi.org/10.1002/app.1988.070350405
14. Holm ER, Kavanagh CJ, Meyer AE, Wiebe D, Nedved BT, Wendt D, Smith CM, Hadfield MG, Swain G, Wood CD, Truby K, Montemarano J (2006) Interspecific variation in patterns of adhesion of marine fouling to silicone surfaces. Biofouling 22(4):233–243. https://doi.org/10.1080/08927010600826129
15. Yesudass SA, Mohanty S, Nayak SK, Rath CC (2017) Zwitterionic–polyurethane coatings for non-specific marine bacterial inhibition: a nontoxic approach for marine application. Eur Polymer J 96:304–315. https://doi.org/10.1016/j.eurpolymj.2017.09.019
16. Ekin A, Webster DC, Daniels JW, Stafslien SJ, Cassé F, Callow JA, Callow ME (2007) Synthesis, formulation, and characterization of siloxane–polyurethane coatings for underwater marine applications using combinatorial high-throughput experimentation. J Coat Technol Res 4:435–451. https://doi.org/10.1007/s11998-007-9039-7
17. Majumdar P, Stafslien S, Daniels J, Webster DC (2007) High throughput combinatorial characterization of thermosetting siloxane–urethane coatings having spontaneously formed microtopographical surfaces. J Coat Technol Res 4:131–138. https://doi.org/10.1007/s11998-007-9015-2
18. Yen MS, Tsai PY (2003) Study on polyethylene glycol/polydimethylsiloxane mixing soft-segment waterborne polyurethane from different mixing processes. J Appl Polym Sci 90(1):233–243. https://doi.org/10.1002/app.12640
19. Wong CS, Badri KH (2012) Chemical analyses of palm kernel oil-based polyurethane prepolymer. https://doi.org/10.4236/msa.2012.32012

20. Queiroz DP, de Pinho MN (2005) Structural characteristics and gas permeation properties of polydimethylsiloxane/poly (propylene oxide) urethane/urea bi-soft segment membranes. Polymer 46(7):2346–2353. https://doi.org/10.1016/j.polymer.2004.12.056
21. Zhou SX, Wu LM, Sun J, Shen WD (2003) Effect of nanosilica on the properties of polyester-based polyurethane. J Appl Polym Sci 88(1):189–193. https://doi.org/10.1002/app.11624
22. Serkis M, Špírková M, Hodan J, Kredatusová J (2016) Nanocomposites made from thermoplastic waterborne polyurethane and colloidal silica. The influence of nanosilica type and amount on the functional properties. Prog Org Coat 101:342–349. https://doi.org/10.1016/j.porgcoat.2016.07.021
23. Ji X, Zhou Y, Zhang B, Hou C, Ma G (2013) Polydimethylsiloxane and castor oil comodified waterborne polyurethane. In: International scholarly research notices, 2013. https://doi.org/10.1155/2013/284683
24. Zhou J, Zhang L, Ma J (2011) Synthesis and properties of nano-SiO2/polysiloxane modified polyacrylate emulsifier-free emulsion. Polym-Plast Technol Eng 50(1):15–19. https://doi.org/10.1080/03602559.2010.512343
25. Mohanty SR, Mohanty S, Samal SK, Nayak SK (2021) Acrylic-ester-polyol based novel two-component polyurethane clear coat: evaluation of performance characteristics. J Appl Polym Sci 138(32):50794. https://doi.org/10.1002/app.50794
26. Pergal MV, Dzunuzović JV, Poreba R, Micić D, Stefanov P, Pezo L, Spirkova M (2013) Surface and thermomechanical characterization of polyurethane networks based on poly (dimethylsiloxane) and hyperbranched polyester. Express Polym Lett 7(10):806–820. https://doi.org/10.3144/expresspolymlett.2013.78
27. Mohanty SR, Mohanty S, Nayak SK, Samal SK (2021) Synthesis and evaluation of novel acrylic and ester-based polyols for transparent polyurethane coating applications. Mater Today Commun 27:102228. https://doi.org/10.1016/j.mtcomm.2021.102228
28. Yesudass SA, Mohanty S, Nayak SK (2018) Facile synthesis of bio-sourced polyurethane-fluorosilane modified TiO 2 hybrid coatings for high-performance self cleaning application. J Polym Res 25:1–10. https://doi.org/10.1007/s10965-017-1430-1
29. Mahoney CM, Gardella JA, Rosenfeld JC (2002) Surface characterization and adhesive properties of poly (imidesiloxane) copolymers containing multiple siloxane segment lengths. Macromolecules 35(13):5256–5266. https://doi.org/10.1021/ma010353y
30. Ramesh S, Mathivanan N (2009) Screening of marine actinomycetes isolated from the Bay of Bengal, India for antimicrobial activity and industrial enzymes. World J Microbiol Biotechnol 25:2103–2111. https://doi.org/10.1007/s11274-009-0113-4

A Multi-parametric Optimization Strategy for Maximizing Bacterial Cellulose Production from Fermented Tea

Baishali Dey, Sivaraman Jayaraman, Bhaskar Das, and Paramasivan Balasubramanian

Abstract Bacterial cellulose (BC) holds great significance as a sustainable raw material in the food, biomedical, cosmetic, electronic, and paper industries. A high concentration of carbon sources and other media components is required to increase the yield of BC, which increases the overall bioprocess cost. Thus, finding alternative substrates and developing optimizing strategies for the cost-effective production of BC becomes indisputable. In this study, the impact of tea leaves and sugar concentration, surface area to depth ratio, scaled-up volume, and different varieties of tea leaves on the production of BC from fermented tea or kombucha (kombucha-derived bacterial cellulose or KBC) has been studied in detail. It was found that 8.0% (w/v) sugar and 0.6% tea leaf concentration gave an optimum KBC productivity of 11.91 ± 0.18 g/L/d. A surface area to depth ratio of 186 cm gave the maximum productivity (29.74 ± 0.65 g/L/d) of KBC. Upscaling of fermenter volume resulted in an approximately two-fold increase in wet weight of KBC. Furthermore, the study also considers the reuse of spent tea leaves and processing waste given environmental sustainability. These outcomes aim to advance the understanding of KBC production, providing insights into maximizing its production for sustainable biomaterial synthesis with diverse applications.

Keywords Bacterial cellulose · Fermented tea · Kombucha · Biomaterial synthesis · Optimization · SCOBY

B. Dey · S. Jayaraman · P. Balasubramanian (✉)
Department of Biotechnology and Medical Engineering, National Institute of Technology Rourkela, Rourkela 769008, India
e-mail: biobala@nitrkl.ac.in

B. Das
KIIT-Technology Business Incubator, KIIT University, Bhubaneswar, Odisha 751024, India

© The Author(s), under exclusive license to Springer Nature Singapore Pte Ltd. 2024

M. Suar et al. (eds.), *Advancements in Polymer Technology*, Springer Proceedings in Materials 57, https://doi.org/10.1007/978-981-97-6333-7_2

1 Introduction

Bacterial cellulose (BC) shares an identical chemical composition $(C_6H_{10}O_5)_n$ with plant-based cellulose, yet it differs in terms of its physico-structural properties, growth conditions, and purity levels. Typically, bacterial genera that are recognized for BC production comprise *Acetobacter* (now reclassified as *Komagataeibacter*), *Aerobacter*, *Achromobacter*, *Agrobacterium*, *Pseudomonas*, *Sarcina*, and *Rhizobium*. Among these, *Komagataeibacter*, particularly the *Komagataeibacter xylinus* strain, stands out as the most frequently cited genus and is widely studied as a model strain for substantial BC production [17]. Significant challenges, including high production costs and low yields, often hinder the widespread utility of BC. The primary commercial roadblock arises from the inefficient conversion of sugars in the media to BC on a large scale. Enhancing the yield of cellulose is an important factor that drives the cost of BC [26]. Moreover, most BC production studies focus on the standard model strain *Komagataeibacter xylinus*, limiting the exploration of other strains or their dynamic interactions in a symbiotic environment [36]. Addressing these challenges requires investigating cost-effective media and alternative microorganism combinations to enhance BC production efficiency.

The above-mentioned challenges can be overcome by utilizing a symbiotic culture of bacteria and yeast (SCOBY), wherein the yeast breaks down the sucrose present in the media into glucose and sucrose, which can be taken up by the bacterial strains [14]. Traditionally, SCOBY has been grown in a fermentation media comprising tea and sucrose, which yields a probiotic beverage (Kombucha) and a cellulose-rich pellicle (KBC) at the end of the fermentation duration [37]. However, only a small part of KBC is used as inoculum for the next batch, and most of it is discarded as waste [8]. But in recent years, KBC has steadily found applicability in different domains, such as the textile industry for sustainable fashion [19], as an adsorbent for heavy metal removal from wastewater [24], as a composite in cathode material for use in lithium-sulfur batteries [18], as a biodegradable packaging material [2], as a wound dressing material [27] to name a few. Thus, BC production through kombucha fermentation presents a promising opportunity for sustainable biomaterial synthesis.

For the production of BC from kombucha fermentation, the most crucial input parameters affecting its production can be identified as the concentration and type of carbon and nitrogen sources and operational parameters such as dimensions of the fermenter vessel, including surface area, depth, and volume. The importance of carbon and nitrogen source selection was published by Yim et al. [40], wherein the authors reported that the efficiency of BC production can be optimized based on the selection of nitrogen and carbon sources in the cultivation medium. The nitrogen source, functioning as a vital nutrient, supports the microorganisms' growth and the development of cellular structures. Previously reported sources for nitrogen include yeast extract, ammonium sulfate [9], casein hydrolysate [20], peptone [41], sodium glutamate, glycine [12], and tea extracts [22], to name a few. Another crucial element for enhancing cell growth and metabolism during BC synthesis is the carbon source. Different carbon sources such as sucrose, date syrup, glucose, mannitol [23], acetone, butanol, and ethanol [16] have been employed in BC production. Tea leaves (*Camellia sinensis*), commonly used in kombucha fermentation, contribute not only to the flavor profile but also serve as a critical substrate for BC synthesis. The concentration of sugar in the fermentation medium plays a pivotal role in providing the necessary carbon source for BC production. This study investigates the requirement to optimize both tea leaves and sugar concentration for maximal BC yield.

Another crucial parameter for KBC production is the fermentation vessel's surface area and depth [29], which significantly modulates oxygen transfer and KBC synthesis. Bhat [6] reported an increase in bacterial cellulose production with an increase in surface area to depth ratio. The scale-up of production volume might introduce certain complexities, necessitating an understanding of the consequential impacts on BC yield. Furthermore, an exploration into the utilization of cellulose pellicle as an inoculum is undertaken to elucidate its effect on BC production.

Lately, there has been a development in alternative substrates for kombucha fermentation by utilizing various plant-based raw materials in the fermentation medium during primary fermentation. Examples include the use of *Brassica tournefortii* leaves [30], coffee [39], fruit juices [43], blanched spinach pulp, rooibos, waste such as soybean whey and banana peel extract [11], as well as infusions derived from yerba mate, lavender, green tea, oregano, and fennel [31]. It was further reported that kombucha fermented in yerba mate infusions exhibited elevated cellulose synthesis, showcasing remarkable antioxidant properties in both components—the beverage itself and the resulting cellulose by-product [38]. Thus, an attempt was made to increase the yield of BC production by utilizing different kinds of tea leaves. Furthermore, this research work also explores the recyclability of spent tea leaves and processing waste, supporting the principles of sustainable production. Tea processing or manufacturing waste (TPW), also known as tea waste, primarily results from the black tea manufacturing process. In contrast, spent tea leaves (STLs) refer to the solid remains remaining after brewing [15]. Most tea waste is currently managed through composting, incineration, or landfilling, which harms the environment due to slow degradation and loss of beneficial components [42]. Continuous disposition of STLs in open areas can worsen soil and water pollution and create unpleasant surroundings [25]. Investigating the potential reuse of these by-products not only addresses environmental concerns but also contributes to the economic viability of BC production.

The novelty lies in the comprehensive strategy undertaken in this work, which considers multiple influencing factors to advance bacterial cellulose production and emphasizes sustainability by utilizing black tea processing waste and checking its reusability. Therefore, the main objectives of this study were to enhance the production of BC from kombucha fermentation by optimizing sucrose and tea concentration, surface area to depth ratio, and upscaled volumes, along with the addition of SCOBY as inoculum and different varieties of tea leaves. Further, an attempt was made to check the reusability of spent tea waste for the production of KBC. The outcomes of this study will provide valuable insights into maximizing bacterial cellulose production from kombucha, paving the way for sustainable and economically feasible production of bacterial cellulose for diverse applications in the biomaterial field.

2 Materials and Methods

2.1 Preparation of Kombucha Bacterial Cellulose from Tea Infusion

A sweetened tea mixture was prepared in a sterilized vessel by boiling 80 g of sucrose (commercial grade) in 1 L of purified water. Subsequently, 6 g of black tea processing waste (obtained from Parry Agro Industries Limited, Valparai, Tamil Nadu, India) was steeped in the solution for 5–7 min. Once the tea cooled to room temperature, a 10% inoculum broth containing a predominant symbiotic culture of *Komagataeibacter rhaeticus* and *Brettanomyces bruxellensis* was added to a sterilized fermentation vessel. The vessel was covered with a clean and disinfected muslin cloth, securely fastened with thread, and left undisturbed for 10–14 days at room temperature for static batch fermentation. Following fermentation, the sweetened tea mixture transformed into kombucha and KBC floating on the surface.

2.2 Effect of Different Sugar and Tea Concentrations on the Productivity of Kombucha Bacterial Cellulose

One variable at a time (OVAT) optimization was carried out by utilizing varying concentrations of sucrose (2.0–14.0% (w/v)) and black tea processing waste (0.3–1.5%), keeping the inoculum of previously fermented liquid tea broth (10.0% v/v) constant, to evaluate its influence on the KBC yield and productivity. The fermentation was allowed to continue for a period of 14 days under aseptic conditions at room temperature. KBC was removed after the completion of the fermentation period and washed and weighed (*Shizmadu*, 0.001 accuracy). Further, the thickness was measured using a digital Vernier caliper (*Zhart*) (±0.02 mm/0.001"(<100 mm)).

2.3 Effect of Surface Area to Depth Ratio on the Yield and Productivity of Kombucha Bacterial Cellulose

The effect of surface area and depth of culture medium on the yield of KBC was studied by preparing tea broth for fermentation in dimensionally varying containers. Tea broth of 500 mL was prepared following the procedure described in Sect. 2.1. Containers of different dimensions used for the study are listed in Table 1 (and

Fig. 1). The KBC produced was collected, washed, and weighed after 14 days of fermentation to examine the effect of surface area and depth of the culture medium on cellulose production.

Table 1 Dimensions of diverse fermentation vessels used in the present study

Sample	Length(cm)	Breadth/Diameter (cm)	Height(cm)	Surface area (cm^2)
A	26.0	22.0	1.0	550.0
B	14.0(U), 11.0(B)	14.0(U), 11.0(B)	3.0	550.0
C	–	12.0(U)[a], 8.0(B)	7.5	550.0
D	–	9.5	9.5	550.0
E	–	11.3(U), 9.2(B)	8.5	550.0
F	–	16.0(U), 8.0(B)	5.0	550.0
G	–	10.7	6.4	550.0
H	–	5.7	25.0	550.0

[a]U: Upper surface, B: Bottom surface, Height: Height of fermentation broth

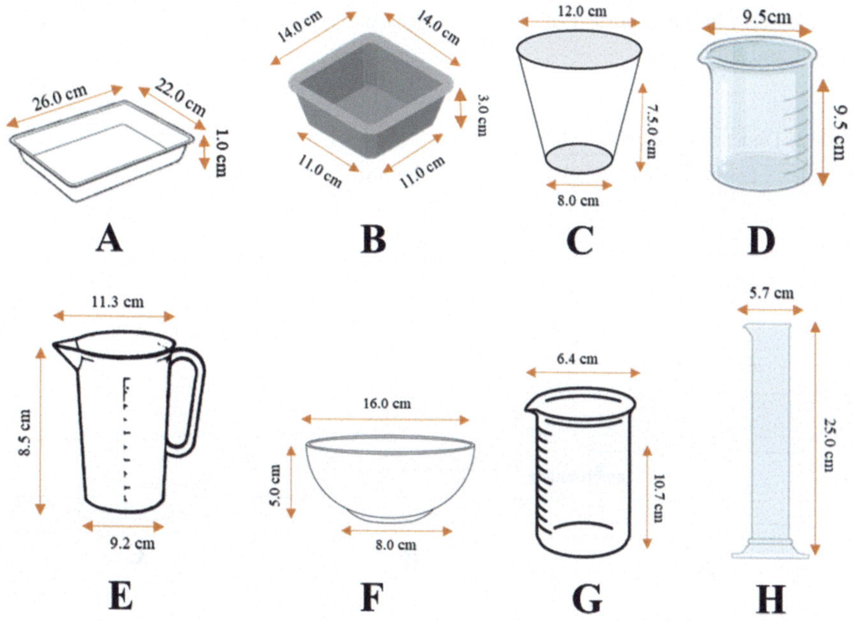

Fig. 1 Schematic representation of different fermentation vessels varying in surface area to depth ratio

2.4 *Effect of Volume of the Fermentation Vessel and Addition of SCOBY Pellicle as Inoculum on the Yield and Productivity of KBC*

To study the effect of volume on the yield and productivity of KBC, three fermentation vessels of increasing volume with the dimensions (Table 2, Fig. 2) were selected. The tea broth was prepared as described in Sect. 2.1 for a volume of 2.0 L, 7.0 L, and 15.0 L. The fermentation broth reached a height of 5.2 cm, 5.6 cm, and 2.5 cm, for 15.0 L, 7.0 L, and 2.0 L, respectively. The open surface of the fermentation vessels was sealed with muslin cloth, and the tea broth was allowed to ferment under static batch conditions for a prolonged period of approximately seven weeks. Every 7th day, approximately 10.0 ml of the sample was collected aseptically for biochemical analysis. After the end of the fermentation duration, the sheets were collected, washed, and measured for weight and thickness.

To study the effect of SCOBY pellicle as inoculum along with inoculum broth, a pellicle of 5.0% (w/v) was added to each of the fermentation vessels along with 10.0% (v/v) of fermentation broth. Another batch was prepared without the addition of SCOBY, and only 10.0% broth was added as inoculum. Static batch fermentation was carried out for a period of 7 weeks, after which the sheets were measured for weight and thickness.

Table 2 Dimensions of up-scaled fermentation vessels used in the present study

Sample	Length (cm)	Breadth (cm)	Height (cm)	Surface area (cm^2)	Volume (L)
A	76.20	37.80	20.50	6946.32	15.00
B	41.20	30.40	15.50	3306.88	7.00
C	35.00	28.00	2.50	2275.00	2.00

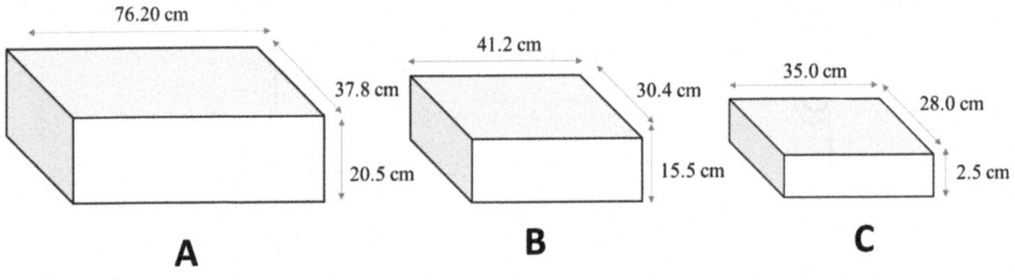

Fig. 2 Schematic representation of scaled up volumes of fermentation vessels

2.5 Effect of Different Tea Leaves on the Productivity of Kombucha Bacterial Cellulose

For studying the effect of different tea leaves on the productivity of KBC, five different kinds of tea leaves were taken for the study, namely- Green tea (GT), Assam Black tea (BT), Black tea variant 1 (BT1), Black tea variant 2 (BT2) and tea processing waste (TPW). BT1 additionally contained *Zingiber officinale* (Ginger, 1.5% w/w), *Elettaria cardamomum* (Cardamom, 1.5% w/w), *Ocimum tenuiflorum* (Tulsi, 0.5% w/w), *Glycyrrhiza glabra* (Licorice 0.5% w/w), and *Withania somnifera* (Ashwagandha, 0.5% w/w), whereas BT2 contained *Zingiber officinale* (Ginger 0.4% w/w), *Syzygium aromaticum* (Clove, 2.0% w/w), *Piper nigrum* (Black Pepper, 2.0% w/w), *Cinnamomum verum* (Cinnamon, 4.0% w/w), *and Elettaria cardamomum* (Cardamom, 2.0% w/w). The tea infusion was prepared according to Sect. 2.1 with different tea leaves and allowed to ferment. Every 7th day, around 10 ml of the sample was collected aseptically for biochemical analysis. Every 14th day, the sheets were collected, washed, and measured.

2.6 Biochemical Changes Occurring in the Fermentation Broth

2.6.1 pH

The pH of the samples was checked daily every 7th or 14th day with a digital pH meter (*Systronics*, MK VI). About 5 ml of the sample was withdrawn carefully without disrupting the cellulose layer to take the measurement.

2.6.2 BRIX

About 0.5 ml of sample was collected using an aseptic dropper every 7th or 14th day to estimate the carbohydrate content using a refractometer (*Hanna,* HI96813, Sugar in wine refractometer; sugar content Range: 0 to 50% Brix; 0.0–25.0% V/V Potential Alcohol) in terms of °BRIX.

2.6.3 TDS

A sample of about 5 ml was collected every 7th or 14th day to measure the Total Dissolved Solids using a TDS pen (*Acu check*, 0–9999 ppm).

2.7 Recyclability of Spent Tea Leaves and Black Tea Processing Waste on the Productivity of KBC

The mixture of spent tea leaves (STL) from the study conducted in Sect. 2.5 and the spent tea processing waste (STPW) from the previous batch experiments (Sects. 2.3 and 2.4) was collected and dried separately in a hot air oven at 30 °C. These dried tea leaves were used for subsequent brewing cycles. The tea infusion was prepared as previously instructed in Sect. 2.1. The sterile beakers were covered with clean and dry muslin cloth, secured with a thread, and left undisturbed for fermentation. After 14 days, the sheets were collected, washed, and measured.

All the experiments were performed in triplicates, and the standard deviation was calculated for every mean value.

3 Results and Discussion

3.1 Production and Optimization of Kombucha Derived Bacterial Cellulose

Keeping the initial operating conditions of inoculum kombucha tea at 10.0% v/v, waste tea leaves of 0.6% w/v, and pH between 3.8 and 4 as constant, OVAT was carried out with varying sugar concentrations from 2.0 to 14.0% with a working volume of 2 L over 14 days to study its effect on KBC yield, thickness and productivity. It was observed that at 8.0% (w/v) sucrose concentration, the maximum productivity of 11 ± 0.05 g/L/d with a thickness of 4.33 ± 0.06 mm (Fig. 3a) was achieved; however, both these aspects decreased with further increase in sucrose (common sugar) concentration. This could be corroborated by the fact that with a further increase in sucrose concentration, a decline in cellulose productivity occurs due to the metabolic conversion of excess sucrose into other by-products [3]. Other possible reasons for the decrease in yield and productivity of cellulose that have been reported are as follows–(1) The higher sugar content in the tea broth during fermentation may increase metabolic product production, potentially leading to product inhibition (2) Uneven rates of transport for essential cell materials and nutrient utilization (3) Intensification of inhibitions to bacterial cellulose synthesis as maintenance of balance between removing harmful substances and bacterial cellulose production is hindered [6] (4) An excessive quantity of sugar extends the lag phase, negatively impacting cellulose production by diminishing cell growth [4] (5) Increased gluconic acid production due to higher metabolic activity hinders bacteria from synthesizing cellulose [5].

Further, keeping the sugar concentration at 8.0% (w/v), with varied tea concentration from 0.3 to 1.5% (w/v), a maximum thickness of 3.48 ± 0.52 mm, wet weight of 333.5 ± 4.95 g, and productivity of 11.91 ± 0.18 g/L/d (Fig. 3b) was obtained at 0.6%

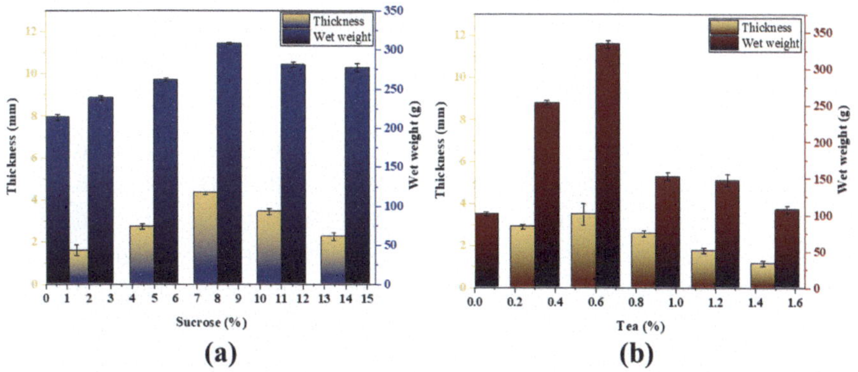

Fig. 3 Change in wet weight, thickness, and productivity of kombucha cellulosic sheet with varying **a** Sugar concentration (%w/v) **b** tea concentration (%w/v)

(w/v). Similar results were also reported by AL-Khalifawi [3], where an enhancement in yield was evident with an increase in tea concentration until a threshold value, beyond which a decline was observed. Since the nitrogenous components and polyphenols play an essential role in governing the metabolism, inhibition of bacterial bio-activity beyond the optimal tea concentration declines the yield. Further, tea contains a high concentration of polyphenols, which have been reported to show antioxidant and antibacterial properties. Thus, at higher tea concentrations, the yield and productivity are decreased [36]. Serra et al. [35] also reported that an increase in tea concentration correlates with a reduction in cellulose yield, as the elevated polyphenol content acts as an inhibitor for biofilm production by gram-negative bacteria. Saavedra-Sanabria et al. [34] published that pH, sugar, and appropriate nitrogenous compounds influence bacterial cellulose biosynthesis. Kurtzman et al. [21] reported that concentrations of tea exceeding 6 g/L may hinder the growth of acetic acid bacteria, consequently resulting in the inhibition of cellulose production.

3.2 *Effect of Surface Area to Depth Ratio on the Yield and Productivity of Kombucha Bacterial Cellulose*

As can be seen in Table 3, a ratio between surface area and depth of 186.00 cm gave the maximum productivity of 29.74 ± 0.65 g/L/d with a depth of 3 cm. It can also be observed that not only the surface area but also the depth of the fermentation vessel plays a pivotal role, which has also been reported by other studies [1, 33]. Less depth results in faster evaporation of the substrate, whereas more depth results in less oxygen availability. Thus, with increasing depth and decreasing surface area to depth ratio, a reduction in the overall yield and KBC productivity was observed. Within a static culture, the BC pellicle is formed at the air–liquid interface, wherein aerobic bacteria exhibit an inclination for cellulose synthesis in proximity to the

Table 3 Change in yield and productivity due to change in surface area to depth ratio

Surface area (cm^2)	Depth (cm)	Ratio surface area: depth (cm)	Wet weight (g)	Yield (g/L)	Productivity (g/L/d)
1240.00	1.00	1240.00	102.93 ± 4.04	205.86 ± 8.08	14.70 ± 0.58
560.00	3.00	186.00	208.20 ± 4.59	416.40 ± 9.18	29.74 ± 0.65
800.63	5.00	160.00	171.87 ± 2.51	343.74 ± 5.02	24.55 ± 0.35
505.00	5.50	91.18	139.30 ± 3.47	278.60 ± 6.95	19.90 ± 0.49
394.97	6.40	61.71	98.46 ± 4.69	196.91 ± 9.39	14.06 ± 0.67
407.21	7.50	54.90	84.14 ± 3.27	168.28 ± 6.55	12.02 ± 0.47
442.55	8.50	52.06	75.92 ± 2.59	151.85 ± 5.18	10.84 ± 0.37
425.29	9.50	44.73	64.74 ± 1.36	129.49 ± 2.73	9.24 ± 0.19
102.07	25.00	4.08	13.15 ± 2.27	26.30 ± 4.55	1.87 ± 0.32

surface [10]. Initially, bacterial proliferation is facilitated through the utilization of dissolved oxygen, resulting in the simultaneous occurrence of cellulose production and increased turbidity. Upon the depletion of dissolved oxygen, the surface bacteria sustain their activity and continue cellulose production. This outcome is rationalized by the fact that cells release carbon dioxide, which becomes trapped in the pellicle, indicating that a container with substantial depth yields a limited quantity of BC. The accumulation of carbon dioxide is heightened in deeper culture mediums. As the interior of the pellicle becomes less aerated, it hinders cell growth and the formation of the pellicle due to the obligate aerobic nature of acetic acid bacteria [3]. Thus, an increase in the surface area of the fermentation medium is crucial for increased cellulose production [6, 17].

3.3 *Effect of Volume of Fermentation Vessel and Addition of SCOBY Pellicle as Inoculum on the Yield and Productivity of KBC*

With an increasing volume of fermentation broth, an increase in cellulose yield and productivity was observed (Table 4). The highest volume of 15.0 L and in the absence of SCOBY pellicle as inoculum yielded the maximum productivity of 9.21 ± 0.02 g/L/d and maximum thickness of 2.01 ± 0.11 mm. When the variation in the depth of the culture medium was minimal, the primary determinant of bacterial cellulose production was the volume of the medium. Under these conditions, cultures with a greater volume of culture medium exhibited higher bacterial cellulose yields [3].

Static cultures with an expansive surface area facilitate cellulose production and mitigate shear stress on cells during the exportation of nanofibers. The surface area is intricately connected to the accessibility of oxygen in the growth medium, primarily

Table 4 Increase in yield and productivity of KBC with increase in volume

Sample	Without SCOBY				With SCOBY			
	Wet weight (g)	Thickness (mm)	Yield (g/L)	Productivity (g/L/d)	Wet weight (g)	Thickness (mm)	Yield (g/L)	Productivity (g/L/d)
A (15.0L)	6775.66 ± 18.14	2.01 ± 0.11	451.71 ± 1.21	9.21 ± 0.02	2011.66 ± 7.50	0.74 ± 0.09	133.87 ± 0.38	2.73 ± 0.01
B (7.0L)	2031.67 ± 17.38	1.52 ± 0.09	290.23 ± 2.48	5.92 ± 0.05	1205.67 ± 8.02	0.82 ± 0.07	172.23 ± 1.14	3.51 ± 0.02
C (2.0L)	175.66 ± 12.22	0.63 ± 0.46	87.83 ± 6.11	2.50 ± 0.17	67.66 ± 4.16	0.92 ± 0.03	33.83 ± 2.08	0.96 ± 0.05

in the form of dissolved oxygen [7]. The aeration state, as dictated by the surface-to-volume (S/V) ratio, constitutes a crucial determinant influencing the synthesis of bacterial cellulose. A greater culture area ensures improved surface aeration, consequently elevating the levels of dissolved oxygen in the broth, thereby enhancing metabolic activity in the SCOBY [38]. It has been further documented that the optimal surface-to-volume ratio for BC production is a characteristic dependent on the strain and is of vital importance. Deviations towards higher or lower ratios have been observed to significantly reduce both the thickness and yield of the BC membrane [10].

3.4 *Effect of Different Tea Leaves on the Productivity of Kombucha Bacterial Cellulose*

In Fig. 4, it can be observed that black tea variant 1 gave the maximum wet weight (1432.67 $\pm$ 18.50 g) and productivity (25.58 $\pm$ 0.28 g/L/d) followed by black tea (1142.33 $\pm$ 20.66 g; 20.39 $\pm$ 0.19 g/L/d). This increase in the wet weight of cellulose can be attributed to the presence of Tulsi, Mulethi, and Ashwagandha in black tea variant 1, whereas in the case of black tea variant 2, a decrease in wet weight was observed, possibly due to the presence of Clove, Black Pepper, and Cinnamon which may have interfered with the growth of the bacterial population. Ginger and cardamom were the common herbal agents present in both the tea leaf varieties and hence can be excluded as the influencing factors in cellulose production. Ramírez Tapias et al. [31] reported that yerba mate, followed by black tea fermentation, showed the highest cellulose production, driven by a greater microbial population. Consortium dynamics and herbal infusions containing cellulose-stimulating xanthines, like theaflavine and caffeine, influenced microbial concentration and film production. However, they also reported that various compounds exhibit antimicrobial properties that may hinder the reproduction of acetic acid bacteria (AAB) and the synthesis of cellulose. For instance, antimicrobial activity was observed in carvacrol and thymol from oregano, linalool from lavender, and neryl acetate and anethole in fennel. As the exact mechanism for the observed variation in cellulose production due to the presence of the aforementioned herbal agents has not been reported yet, it can be explored further in future studies.

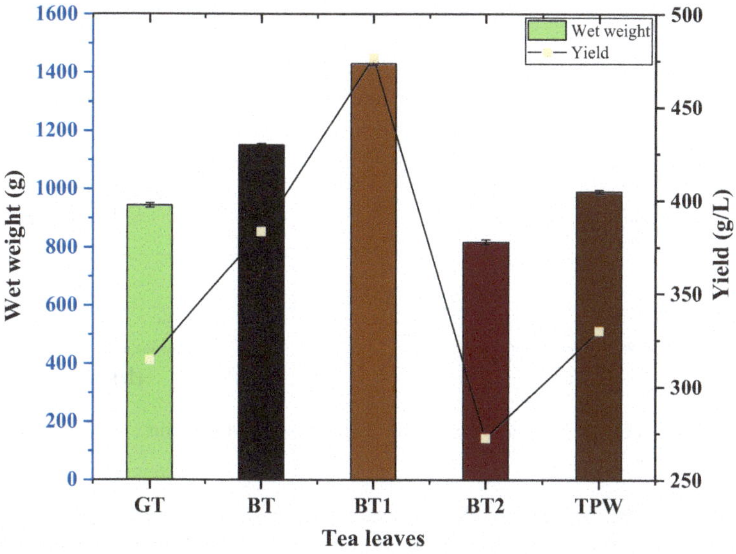

Fig. 4 Impact of different tea leaves on the wet weight and yield of cellulose

3.5 Biochemical Changes Occurring in the Fermentation Broth

3.5.1 pH

In the absence of SCOBY pellicle, a minimum pH of 2.35 ± 0.07, 2.30 ± 0.04 and 2.41 ± 0.10 was achieved, whereas in the presence of SCOBY pellicle, a minimum pH of 2.32 ± 0.04, 2.14 ± 0.06 and 2.38 ± 0.08 was achieved by the end of the fermentation period respectively in decreasing order of the different volumes of fermentation vessels (Fig. 5a). When different types of tea leaves were utilized, different pH values of 2.85 ± 0.06, 2.63 ± 0.05, 2.75 ± 0.06, 2.71 ± 0.19, 2.75 ± 0.10 were obtained by the end of fermentation duration for green tea, black tea, black tea variant 1, and 2 and black tea processing waste respectively (Fig. 5b). As fermentation progresses, the pH of the medium continues to decrease, signaling the occurrence of the fermentation process and the generation of organic acids [29]. The membrane-anchored glucose dehydrogenase (GDH) situated in the bacterial cell membrane transforms glucose into gluconic acid. The build-up of gluconic acid in the culture medium leads to a pH drop, subsequently reducing cell viability and impacting cellulose formation [4]. The fermentation process results in the synthesis of ethanol through the ideation of glucose by the yeast partner in the fermented tea. Subsequently, acetic acid bacteria (AAB) utilize this ethanol to produce acetic acid in the fermented tea medium. The production of acetic acid, in turn, stimulates further ethanol formation, thereby enhancing the metabolic-level production of bacterial

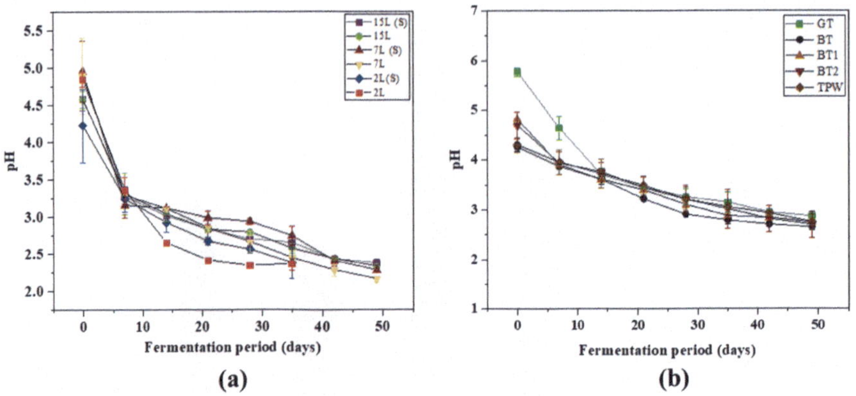

Fig. 5 Decline in pH of fermentation broth over the fermentation duration

nanocellulose [36]. While kombucha tea does contain a microbial population, its primary function is to lower the pH to a mildly acidic level through organic acids. This effectively prevents contamination by pathogenic bacteria and fungi, ensuring the survival of the microbial population in the SCOBY. Nonetheless, the regular monitoring of pH also revealed that variation in different tea leaves or the addition of cellulose as inoculum did not have much effect on the end values of pH. However, the higher volume of fermentation broth may have resulted in the accumulation of more acid during the fermentation period, resulting in lower pH values.

3.5.2 BRIX

In the absence of SCOBY pellicle, the total sugar content reached values of 6.2 $\pm$ 0.17, 6.10 $\pm$ 0.10 and 5.27 $\pm$ 2.14, whereas in the presence of SCOBY pellicle, values of 6.17 $\pm$ 0.12, 6.13 $\pm$ 0.12 and 6.93 $\pm$ 0.15 were achieved by the end of fermentation period respectively in decreasing order of the different volumes of fermentation vessels (Fig. 6a). When different types of tea leaves were utilized, sugar content decreased up to 7.40 $\pm$ 0.07, 5.93 $\pm$ 0.15, 5.90 $\pm$ 0.10, 5.95 $\pm$ 0.15, 6.13 $\pm$ 0.06 by the end of fermentation duration for green tea, black tea, black tea variant 1 and 2 and black tea processing waste respectively (Fig. 6b). The initial rise in total sugar content is because the yeast breaks down the sucrose into glucose and fructose. This glucose is then consumed by the bacteria, which results in a decrease in total sugar content [28]. Forsvall [13] also reported a reduction in sucrose content, accompanied by an increase in glucose and fructose content indicated by the sugar analysis during the fermentation period. In the case of green tea, the slow increase in sucrose content may be attributed to the antibacterial potential of green tea compared to its counterparts, which might have delayed the growth of the bacterial strain and also led to a decrease in cellulose production.

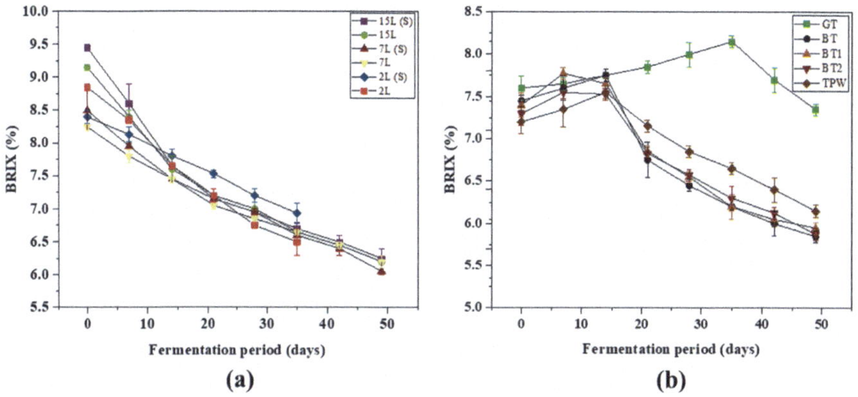

Fig. 6 Change in total sugar content during the fermentation period

3.5.3 TDS

In the absence of SCOBY pellicle, the total dissolved solids content reached values of 881.23 ± 9.29 ppm, 939.67 ± 4.04 ppm, and 965.66 ± 211.50 ppm, whereas in the presence of SCOBY pellicle, values of 1006.67 ± 12.58 ppm, 936.33 ± 8.08 ppm, and 982.33 ± 9.07 ppm were achieved by the end of the fermentation period, respectively, in decreasing order of the different volumes of fermentation vessels (Fig. 7a). When different types of tea leaves were utilized, TDS values increased up to 384 ± 4.51 ppm, 405.33 ± 3.72 ppm, 506.00 ± 6.56 ppm, 564.00 ± 9.64 ppm, 489.00 ± 7.93 ppm by the end of fermentation duration for green tea, black tea, black tea variant 1 and 2 and black tea processing waste respectively (Fig. 7b). The increase in total dissolved solids of the fermentation broth was observed to increase during the course of fermentation, which can be a qualitative indicator of the rise in bioactive components released into the fermentation broth not only by microbial metabolism but also from the tea leaves (such as polyphenols, caffeine, amino acids). Razmovski and Vucurovic [32] also reported that the concentration of TDS in all fermentation systems exhibited a gradual and linear increase over the fermentation period, attributed to the metabolic activity of yeast and the release of cell contents from yeast cells.

3.6 Recyclability of Black Tea Processing Waste on the Productivity of KBC

The use of black tea processing waste and further reuse of spent tea leaves add towards achieving Sustainable Development Goals (SDGs). It was observed that (Table 5), though, in the first brewing cycle, STPW produced a higher wet weight of cellulose as compared to the STL; in the subsequent brews, STPW showed a higher decline as

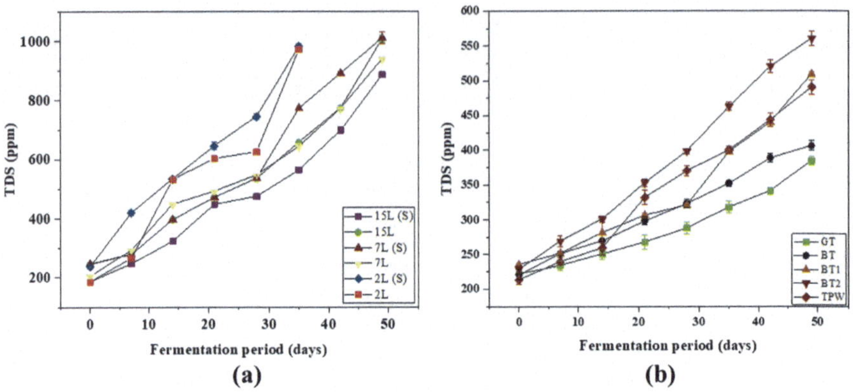

Fig. 7 Increase in concentration of Total dissolved solids during fermentation

compared to STL. After five brewing cycles, the wet weight of KBC reduced from 37.66 ± 1.05 g to 6.96 ± 0.71 g for STL, whereas for STPW, it reduced from 55.07 ± 3.12 g to 5.28 ± 0.23 g respectively. Despite its generation as waste, spent tea leaves contain polyphenols, alkaloids, essential amino acids, fatty acids, and minerals and can serve as an effective carbon source for fermentation after pre-treatment [25]. In the study by [4], the production of cellulose reached 18.68 g/L and 10.6 g/L when utilizing the first brewed black tea and its tenth brew, respectively. Despite re-brewing the black tea waste ten times, only a 43.25% decrease in cellulose production (10.6 g/L) was observed. The variation in the percentage reduction of cellulose production may be due to the difference in the quality of tea leaves, fermentation conditions, and the inoculum culture used. [36] achieved productivity of 12.8 g/L of cellulose from used tea leaves on a dry weight basis after 20 days at 60 g/L of sucrose concentration. Thus, even though the reusability of STL and STPW may not hold much promise, it can still act as a no-cost medium for cellulose production.

Table 5 Average wet weight of KBC and its percentage reduction for spent tea leaves waste and spent tea processing waste

Sample	Average wet weight (g)		Reduction (%)	
	STL	STPW	STL	STPW
1 (1st brew)	37.66 ± 1.05	55.07 ± 3.12	–	–
2 (2nd brew)	8.83 ± 2.85	7.45 ± 0.93	76.55	86.47
3 (3rd brew)	6.67 ± 1.03	6.71 ± 1.24	82.28	87.81
4 (4th brew)	6.64 ± 1.12	6.10 ± 0.54	82.36	88.92
5 (5th brew)	6.91 ± 0.95	5.49 ± 0.29	81.65	90.03
6 (6th brew)	6.96 ± 0.71	5.28 ± 0.23	81.52	90.41

4 Conclusion

Thus, from this study, it can be seen that various factors play into the production of bacterial cellulose from kombucha, which can be optimized to maximize its production. The sugar and tea concentration was optimized at 8.0% (w/v), and 0.6% (w/v), respectively, which gave a maximum productivity of 11.91 ± 0.18 g/L/d. Further, the importance of oxygen availability was determined by observing the impact of surface area to depth ratio and upscaled volume on the productivity and yield of cellulose. The presence of herb infusion in tea leaves was seen to enhance cellulose production. It was also observed that tea processing waste could also serve as a potential nitrogenous source for the economical production of kombucha-derived bacterial cellulose. Though the recyclability of spent tea leaves saw a sharp decline in average wet weight, it could still be utilized in combination with fresh tea leaves for the sustainable production of bacterial cellulose. Thus, using a symbiotic culture of *Komagataeibacter rhaeticus* and *Brettanomyces bruxellensis* to ferment black tea processing waste for the enhanced production of bacterial cellulose can prove to be an economical and sustainable method for different applications.

Data availability

All the data supporting the findings of the study are available with the manuscript and its supplementary material.

Ethics approval

This declaration is not applicable.

Statements and Declarations

No funding was received to assist with the preparation of this manuscript.

The authors have no relevant financial or non-financial interests to disclose.

Author contribution

All authors contributed to the study's conception and design. Material preparation, data collection, and analysis were performed by Baishali Dey. The first draft of the manuscript was written by Baishali Dey, and all authors commented on previous versions of the manuscript. All authors read and approved the final manuscript.

Acknowledgements The authors are grateful to the Department of Biotechnology and Medical Engineering, National Institute of Technology Rourkela, for providing their research facilities. The authors acknowledge the Ministry of Education, Government of India, for sponsoring the first author's PhD program.

References

1. Abd El-Salam SS (2012) Bacterial cellulose of kombucha mushroom tea. N Y Sci J 5(4). http://www.sciencepub.net/newyork
2. Aduri P, Ankita Rao K, Fatima A, Kaul P, Shalini A (2019) Study of biodegradable packaging material produced from SCOBY. J Life Sci Bioinform Pharm Chem Sci 5(3):389–404. https://doi.org/10.26479/2019.0503.32
3. AL-Kalifawi EJ, Hassan IA (2014) Factors Influence the yield of bacterial cellulose of Kombucha (Khubdat Humza). Baghdad Sci J 11(3):1420–1428
4. Avcioglu NH, Birben M, Seyis Bilkay I (2021) Optimization and physicochemical characterization of enhanced microbial cellulose production with a new Kombucha consortium. Process Biochem 108:60–68. https://doi.org/10.1016/J.PROCBIO.2021.06.005
5. Balasubramanian P, Praharaj PT (2023) Principal component analysis revealed the key influencing factors of kombucha bacterial cellulose yield and productivity. Bioresour Technol Rep 23:101539. https://doi.org/10.1016/J.BITEB.2023.101539
6. Bhat, R. (2012). Fermentation of black tea broth (Kombucha): I. Effects of sucrose concentration and fermentation time on the yield of microbial cellulose. Int Food Res J 19(1):109–117
7. Blanco Parte FG, Santoso SP, Chou CC, Verma V, Wang HT, Ismadji S, Cheng KC (2020) Current progress on the production, modification, and applications of bacterial cellulose. Crit Rev Biotechnol 40(3):397–414. https://doi.org/10.1080/07388551.2020.1713721
8. Cubas ALV, Provin AP, Dutra ARA, Mouro C, Gouveia IC (2023) Advances in the production of biomaterials through kombucha using food waste: concepts, challenges, and potential. Polymers 15(7):1701. https://doi.org/10.3390/polym15071701. (Multidisciplinary Digital Publishing Institute)
9. Dirpan A, Kamaruddin I, Syarifuddin A, Zainal, Rahman ANF, Hafidzah, Latief R, Prahesti KI (2019) Characteristics of bacterial cellulose derived from two nitrogen sources: ammonium sulfate and yeast extract as an indicator of smart packaging on fresh meat. IOP Conf Ser Earth Environ Sci 355(1):012040. https://doi.org/10.1088/1755-1315/355/1/012040
10. El-Gendi H, Taha TH, Ray JB, Saleh AK (2022) Recent advances in bacterial cellulose: a low-cost effective production media, optimization strategies, and applications. Cellulose 29(14):7495–7533. https://doi.org/10.1007/s10570-022-04697-1. (Springer)
11. Emiljanowicz KE, Malinowska-Pańczyk E (2020) Kombucha from alternative raw materials–the review. Crit Rev Food Sci Nutr 60(19):3185–3194. https://doi.org/10.1080/10408398.2019.1679714
12. de Fernandes IAA, Pedro AC, Ribeiro VR, Bortolini DG, Ozaki MSC, Maciel GM, Haminiuk CWI (2020) Bacterial cellulose: from production optimization to new applications. Int J Biol Macromol 164:2598–2611. https://doi.org/10.1016/j.ijbiomac.2020.07.255. (Elsevier)
13. Forsvall J (2021) The effect of temperature and fungal: bacterial ratio on kombucha culture fermentation.
14. Geraris Kartelias I, Karantonis HC, Giaouris E, Panagiotakopoulos I, Nasopoulou C (2023) Kombucha fermentation of Olympus mountain tea (Sideritis scardica) sweetened with thyme honey: physicochemical analysis and evaluation of functional properties. Foods 12(18):3496. https://doi.org/10.3390/foods12183496
15. Guo S, Kumar Awasthi M, Wang Y, Xu P (2021) Current understanding in conversion and application of tea waste biomass: a review. Biores Technol 338:125530. https://doi.org/10.1016/J.BIORTECH.2021.125530
16. Huang C, Yang XY, Xiong L, Guo HJ, Luo J, Wang B, Zhang HR, Lin XQ, Chen XD (2015) Evaluating the possibility of using acetone-butanol-ethanol (ABE) fermentation wastewater for bacterial cellulose production by Gluconacetobacter xylinus. Lett Appl Microbiol 60(5):491–496. https://doi.org/10.1111/lam.12396
17. Jozala AF, de Lencastre-Novaes LC, Lopes AM, de Carvalho Santos-Ebinuma V, Mazzola PG, Pessoa-Jr A, Grotto D, Gerenutti M, Chaud MV (2016) Bacterial nanocellulose production and application: a 10-year overview. Appl Microbiol Biotechnol 100(5):2063–2072. https://doi.org/10.1007/s00253-015-7243-4. (Springer)

18. Kalaiappan K, Rengapillai S, Marimuthu S, Murugan R, Thiru P (2020) Kombucha SCOBY-based carbon and graphene oxide wrapped sulfur/polyacrylonitrile as a high-capacity cathode in lithium-sulfur batteries. Front Chem Sci Eng 14(6):976–987. https://doi.org/10.1007/S11705-019-1897-X/METRICS
19. Kamiński K, Jarosz M, Grudzień J, Pawlik J, Zastawnik F, Pandyra P, Kołodziejczyk AM (2020) Hydrogel bacterial cellulose: a path to improved materials for new eco-friendly textiles. Cellulose 27(9):5353–5365. https://doi.org/10.1007/S10570-020-03128-3/FIGURES/7
20. Kolesovs S, Neiberts K, Beluns S, Gaidukovs S, Semjonovs P (2022) Bacterial cellulose production by Novacetimonas hansenii MSCL 1646 on apple juice. Appl Microbiol Biotechnol 106(22):7449–7460. https://doi.org/10.1007/S00253-022-12213-5/TABLES/4
21. Kurtzman CP, Robnett CJ, Basehoar-Powers E (2001) Zygosaccharomyces kombuchaensis, a new ascosporogenous yeast from 'Kombucha tea.' FEMS Yeast Res 1(2):133–138. https://doi.org/10.1111/J.1567-1364.2001.TB00024.X
22. Laavanya D, Shirkole S, Balasubramanian P (2021) Current challenges, applications and future perspectives of SCOBY cellulose of Kombucha fermentation. J Clean Prod 295:126454. https://doi.org/10.1016/j.jclepro.2021.126454. (Elsevier)
23. Mohammadkazemi F, Azin M, Ashori A (2015) Production of bacterial cellulose using different carbon sources and culture media. Carbohyd Polym 117:518–523. https://doi.org/10.1016/J.CARBPOL.2014.10.008
24. Mousavi SM, Hashemi SA, Babapoor A, Savardashtaki A, Esmaeili H, Rahnema Y, Mojoudi F, Bahrani S, Jahandideh S, Asadi M (2019) Separation of Ni (II) from industrial wastewater by Kombucha Scoby as a colony consisted from bacteria and yeast: kinetic and equilibrium studies. Acta Chimica Slovenica, 66(4), 865–873. https://doi.org/10.17344/acsi.2019.4984
25. Negi T, Kumar Y, Sirohi R, Singh S, Tarafdar A, Pareek S, Kumar Awasthi M, Alok Sagar N (2022) Advances in bioconversion of spent tea leaves to value-added products. Bioresour Technol 346:126409. https://doi.org/10.1016/J.BIORTECH.2021.126409
26. Pandey A, Singh A, Singh MK (2023) Novel low-cost green method for production bacterial cellulose. Polym Bull 1–21. https://doi.org/10.1007/S00289-023-05023-W/FIGURES/6
27. Pooja R, Vadodaria K, Vidhya S (2019) Synthesis of bacterial cellulose and herbal extract for the development of wound dressing. Mater Today Proc 15:284–293. https://doi.org/10.1016/J.MATPR.2019.05.007
28. Pradhan S, Prabhakar MR, Karthika Parvathy KR, Dey B, Jayaraman S, Behera B, Paramasivan B (2023) Metagenomic and physicochemical analysis of Kombucha beverage produced from tea waste. J Food Sci Technol 60(3):1088–1096. https://doi.org/10.1007/S13197-022-05476-3/TABLES/2
29. Priyadharshini T, Nageshwari K, Vimaladhasan S, Parag Prakash S, Balasubramanian P (2022) Machine learning prediction of SCOBY cellulose yield from Kombucha tea fermentation. Bioresource Technology Reports 18:101027. https://doi.org/10.1016/J.BITEB.2022.101027
30. Rahmani R, Beaufort S, Villarreal-Soto SA, Taillandier P, Bouajila J, Debouba M (2019) Kombucha fermentation of African mustard (Brassica tournefortii) leaves: chemical composition and bioactivity. Food Biosci 30:100414. https://doi.org/10.1016/J.FBIO.2019.100414
31. Ramírez Tapias YA, Di Monte MV, Peltzer MA, Salvay AG (2022) Bacterial cellulose films production by Kombucha symbiotic community cultured on different herbal infusions. Food Chem 372:131346. https://doi.org/10.1016/j.foodchem.2021.131346
32. Razmovski R, Vučurović V (2012) Bioethanol production from sugar beet molasses and thick juice using Saccharomyces cerevisiae immobilized on maize stem ground tissue. Fuel 92(1):1–8. https://doi.org/10.1016/J.FUEL.2011.07.046
33. Ruka DR, Simon GP, Dean KM (2012) Altering the growth conditions of Gluconacetobacter xylinus to maximize the yield of bacterial cellulose. Carbohyd Polym 89(2):613–622. https://doi.org/10.1016/J.CARBPOL.2012.03.059
34. Saavedra-Sanabria OL, Durán D, Cabezas J, Hernández I, Blanco-Tirado C, Combariza MY (2021) Cellulose biosynthesis using simple sugars available in residual cacao mucilage exudate. Carbohyd Polym 274:118645. https://doi.org/10.1016/J.CARBPOL.2021.118645

35. Serra DO, Mika F, Richter AM, Hengge R (2016) The green tea polyphenol EGCG inhibits E. coli biofilm formation by impairing amyloid curli fibre assembly and downregulating the biofilm regulator CsgD via the σE-dependent sRNA RybB. Mol Microbiol 101(1):136–151. https://doi.org/10.1111/MMI.13379
36. Sharma C, Bhardwaj NK (2019) Biotransformation of fermented black tea into bacterial nanocellulose via symbiotic interplay of microorganisms. Int J Biol Macromol 132:166–177. https://doi.org/10.1016/J.IJBIOMAC.2019.03.202
37. Su J, Tan Q, Tang Q, Tong Z, Yang M (2023) Research progress on alternative kombucha substrate transformation and the resulting active components. Front Microbiol 14. https://doi.org/10.3389/FMICB.2023.1254014
38. Tapias YAR, Di Monte MV, Peltzer MA, Salvay AG (2023) Kombucha fermentation in yerba mate: cellulose production, films formulation and its characterisation. Carbohydr Polym Technol Appl 5:100310. https://doi.org/10.1016/J.CARPTA.2023.100310
39. Watawana MI, Jayawardena N, Waisundara VY (2015) Enhancement of the functional properties of coffee through fermentation by "Tea Fungus" (Kombucha). J Food Process Preserv 39(6):2596–2603. https://doi.org/10.1111/JFPP.12509
40. Yim SM, Song JE, Kim HR (2017) Production and characterization of bacterial cellulose fabrics by nitrogen sources of tea and carbon sources of sugar. Process Biochem 59:26–36. https://doi.org/10.1016/J.PROCBIO.2016.07.001
41. Zhang H, Chen C, Zhu C, Sun D (2016) Production of bacterial cellulose by Acetobacter xylinum: effects of carbon/nitrogen ratio on cell growth and metabolite production. Cellul Chem Technol 50(9)
42. Zhao T, Chen Z, Lin X, Ren Z, Li B, Zhang Y (2018) Preparation and characterization of microcrystalline cellulose (MCC) from tea waste. Carbohyd Polym 184:164–170. https://doi.org/10.1016/J.CARBPOL.2017.12.024
43. Zubaidah E, Dewantari FJ, Novitasari FR, Srianta I, Blanc PJ (2018) Potential of snake fruit (Salacca zalacca (Gaerth.) Voss) for the development of a beverage through fermentation with the Kombucha consortium. Biocatal Agric Biotechnol 13:198–203. https://doi.org/10.1016/J.BCAB.2017.12.012

Analysing Industrial Scale-Up of Carbon Dioxide Capture in Aqueous Amino Acids for Sustainable Technologies

V. Praveenkumar, Harshit Nitin Mittal, and Omkar Singh Kushwaha

Abstract Recent statistics and trends indicate that fossil fuels are the leading cause of carbon dioxide emissions. As a multi-stage supply chain model with a well-organized report on capturing, utilizing, converting, and storing carbon dioxide, CCUS (Carbon Capture, Utilisation, and Storage) is critically assessed. In this paper, we describe using the amino acid in aqueous form as a green, environmentally friendly method of CO_2 capture at ambient temperature (298.15 K) and pressure of around 1.5 Mpa. We studied the process of CO_2 capture in a high-pressure triple injection setup equipped with a data acquisition system to monitor and record input values in real-time. Each experiment was done thrice to determine the saturation (estimate carbon dioxide uptake capacity) and the feasibility of designing a continuous process with minimum percentage error. The experimentation was done in an optimized concentration of amino acids with water in a 100 ml reactor and was also projected for industrial scale-up in a 100 L reactor. To interpret the impact of mass transfer on CO_2 uptake, a common surfactant, sodium dodecyl sulfate (SDS) and water, was utilized, and the results were reported. The capacity of such amino acids is majorly analyzed concerning comparative analysis with other amino acids, amines, amides, etc., for carbon capture.

V. Praveenkumar
Department of Chemical Engineering, Institute of Chemical Technology Mumbai-Indian Oil Campus, Bhubaneswar, Odisha 751013, India

H. N. Mittal
University School of Chemical Technology, Guru Gobind Singh Indraprastha University, Dwarka, Delhi 110078, India

Pro H2Vis Solutions, KLE-CTIE, KLE Technological University, Huballi, Karnataka 580031, India

O. S. Kushwaha (✉)
Chemical Engineering Department, Indian Institute of Technology, Madras, Tamil Nadu, Chennai 600036, India
e-mail: kushwaha.iitmadras@gmail.com

© The Author(s), under exclusive license to Springer Nature Singapore Pte Ltd. 2024

M. Suar et al. (eds.), *Advancements in Polymer Technology*, Springer Proceedings in Materials 57, https://doi.org/10.1007/978-981-97-6333-7_3

Graphical Abstract CCUS overview through L-Histidine Experimental Setup.

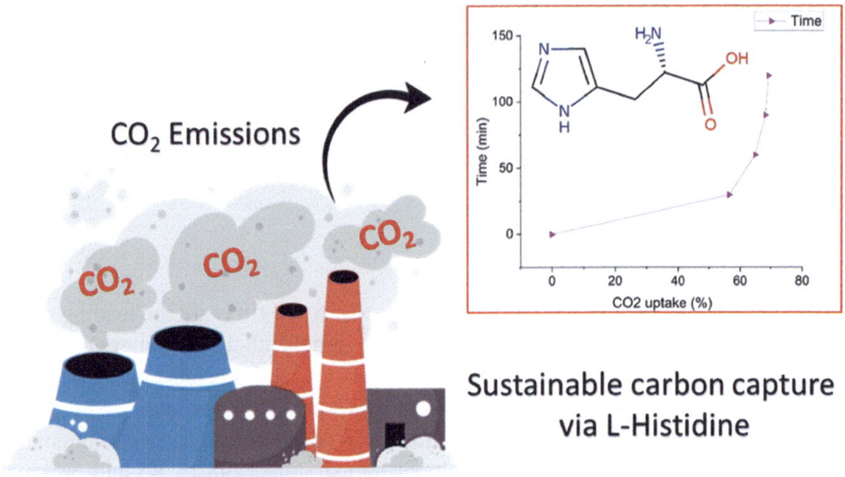

Keywords Carbon capture and utilisation · Emissions · Storage · Industrial Scale-up

1 Introduction

One of the most pressing issues confronting today's globe and generation is the concern about climate change and global warming produced by greenhouse emissions, particularly carbon dioxide emissions, which account for 64% of total emissions from greenhouse gases. According to the UNECE (United Nations Economic Commissions of Europe), to meet the 2050 targets, those countries would need to reduce their dependency on fossil fuels from over 80% to around 50%, as well as their overall carbon emissions. To stay on track to meet the two-degree Celsius target, the UNECE region's nations must reduce or capture at least 90 Gt of CO2 emissions by 2050 [11, 17, 29, 34, 41, 70].

CCUS (Carbon Capture, Utilisation, and Storage) is a supply chain management model that captures carbon dioxide emissions before processing, compressing, and transporting them to be stored or used. The essential components of a CCUS supply chain network are (i) a CO_2 source, (ii) a capture plant, (iii) a transportation pipeline, (iv) a utilization location, and (v) a sequestration site. The CO_2 is separated or captured using a technique at the capture facility. Depending on the separation method used for CO_2 capture, each step employs the material required for the selective separation of CO_2 from flue gas. Once the CO_2 has been removed, it is compressed, and transported to its final destination for utilization or sequestration. A

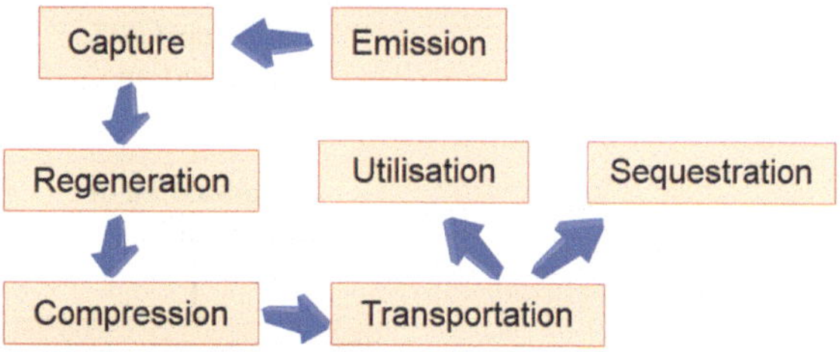

Fig. 1 CCUS as a supply chain model [1, 11, 15]

detailed CCUS supply chain model consisting of emissions, capture, regeneration, compression, transportation, utilization, and sequestration is shown in Fig. 1, [1, 11, 15].

For progress in a specific research field, there should be a thorough report or review with necessary insights, novelties, and new directions for proper commercialization, industrialization and technological advancements [8, 9, 31, 41, 56, 64, 71]. The primary goal of carbon capture, utilization, and storage (CCUS) is to reach net zero emissions as soon as practicable. Regarding current trends, we can see that there have yet to be significant agreements among stakeholders and trustworthy, inexpensive, and more efficient CCUS facilities/models. When the present research on carbon neutrality and carbon dioxide storage was examined later in the review, it was discovered that more research on net zero emissions (NZE) and carbon neutrality was needed before the COVID period (1900s–2020) [9, 40, 42, 48, 55, 65, 79].

1.1 Carbon Capture

Carbon Capture is one method for reducing carbon dioxide emissions by capturing and then storing or utilizing it. Carbon capture can be accomplished by post-combustion, oxygen-combustion, and pre-combustion. It is a technologically significant, critical, long-term step toward renewable energy generation [4, 20, 36, 51, 59, 61, 71]. It absorbs CO_2 emissions before they are released into the atmosphere. However, because carbon capture is expensive, one of the critical obstacles to this technological progress is affordability [66].

Pre-combustion is mainly used in coal-gasification plants, although post-combustion and oxy-fuel combustion are used in coal and gas-fired facilities. When several studies compared all three pathways, it was observed that the pre-combustion technology projected the lowest CPT (cost per ton) of CO_2 for coal-fired plants. However, post-combustion and oxy-fuel technologies showcase similar costs for coal-fired plants. But when the gas-fired plants were considered, the CPT of post-combustion capture was half of the pre-combustion and oxy-combustion technologies. The least efficient pathway for CO_2 capture is post-combustion CO_2 capture, with an energy penalty of approximately 8% and 6% for coal and gas plants, respectively. The overview of pathways of carbon capture is shown in Fig. 2 [7].

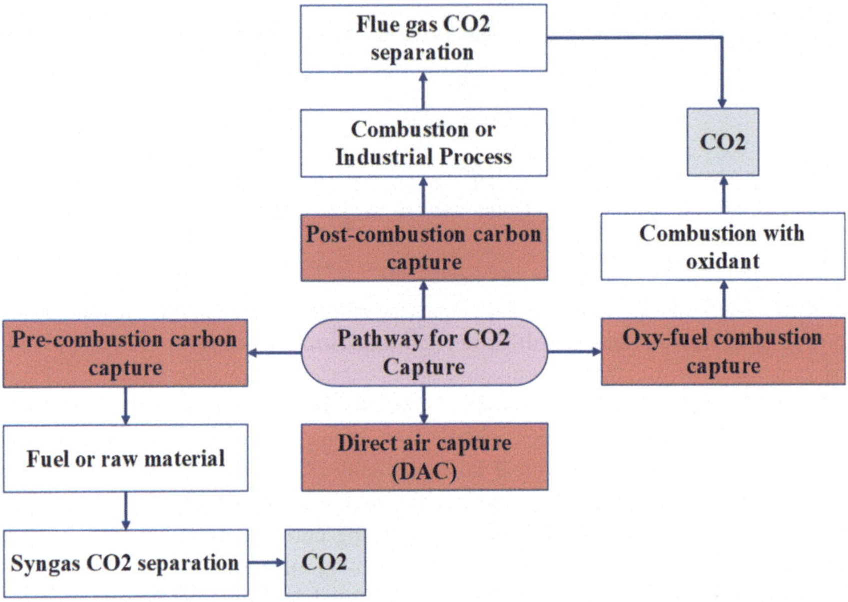

Fig. 2 Outlook of various carbon capture methods [7]

Carbon Capture and Sequestration technologies consist of absorption, adsorption, cryogenics, membranes, microbial, and algal systems [19, 32, 54]. Technologies and materials are essential for such carbon capture technologies to find the best uptake capacity of carbon and adsorption parameters based on low, intermittent, and high carbon industries. Table 1 showcases the carbon capture and sequestration technology, technologies, and materials [37].

1.2 Amines, Ionic Liquids and Synthetic Material for Carbon Capture

A class of substances known as amines and polyamines is up-and-coming for CO_2 capture and is increasingly used in techniques to allow CO_2 utilization. Several variables and performance metrics related to CO_2 capture are crucial for assessing an adsorbent's capacity for separation. The approaches and situations under which the suggested CO_2 adsorbents are studied are also considered. The researcher needs this knowledge to find new carbon capture materials or improve existing ones. Selectivity is a crucial component in assessing the separation capability of an adsorbent. To obtain high product purity, this refers to the adsorbent's preferred affinity towards the desired adsorbate (CO_2, for example) over other components present in the gas mixture, like nitrogen [6, 21, 28, 52, 63, 73, 76].

Table 1 Historical Roadmap of Carbon Capture and Sequestration Technologies, technologies and materials [37]

Carbon capture and sequestration technology	Technologies	Materials	References
Absorption	Chemical	MEA	[27]
		Caustic	[74]
	Physical	Selexol	[58]
		Rectisol	[49]
Adsorption	Adsorber beds	Alumina	[75]
		Zeolite	[22]
		Activated C	[38]
	Regeneration method	Pressure swing	[58]
		Temperature swing	[57]
		Washing	[10]
Cryogenics	Bio-fixation	Enzyme-catalytic hydrolysis	[46]
Membranes	Gas separation	Polyphenylene oxide	[78]
		Polydimethylsiloxane	[13]
	Gas absorption	Polypropylene	[77]
	Ceramic based systems	Lithium based Ceramics	[33]
Microbial/algal systems	Conventional V-L	Cryogenic liquid	[53]

Some significant factors that affect CO_2 adsorption involve temperature, humidity, design of CO_2 adsorbents, CO_2 selectivity, and stability (thermal stability, oxidative degradation, CO_2-induced degradation, Poisoning by Sox and NOx) [2, 44, 47]. Scale for carbon capture technologies, especially from the industrial perspective for various amines and ionic liquids, includes selectivity, high-throughput screening of materials, process modeling and simulation, and realistic commercial ventures in carbon capture technologies for amino acids, amines, and ionic liquids. Many carbon capture materials are showcased in Table 2.

High CO_2 selectivity over other components, namely N_2, is required for CO_2 capture from either air or a static point. However, because of its non-polarity and dearth of advantageous adsorption sites, activated carbon often does not have a strong selectivity for CO_2 over N_2. A polar surface with a large number of micropores (pore diameters typically less than 0.5 nm) may enhance CO_2 selectivity over N_2, taking into account CO_2's electric quadrupole moment and the small dynamic diameter difference between N_2 and CO_2. Whereas when we discuss CO_2 selectivity among different capture materials, such things can only be analyzed if the carbon capacity occurs at similar thermodynamic as well as exergy parameters. If different capture materials are handled at different parameters, then it becomes difficult to homogenize and standardize those materials [18, 35, 39, 68].

Table 2 Classification of conventional carbon capture materials

Carbon capture materials		References
Amines	Monoethanolamine	[16]
	Diethanolamine	[5]
	Methyldiethanolamine	[60]
	Aminomethyl propanolamine	[30]
	Piperazine	[24]
	Diglycolamine	[67]
	Diisopropanolamine	[76]
Ionic liquids	HEF	[72]
	THEAA	[26]
	HEAF	[50]
	HEAA	[12]
	TEGO IL K5	[62]
Synthetic materials	CaO	[23]
	$Ca_{12}Al_{14}O_{33}$	[23]
	MgO	[25]

1.3 Environmental and Economic Impacts of Amines, Ionic Liquids, Synthetic Materials and Amino Acids

In Carbon Capture technologies, especially for an industrial scale-up, amino acids like L-valine and L-histidine are preferred over traditional amines due to their resistance to degradation and deterioration in oxygen-rich flue gas and better CO_2 absorption. An additional benefit is their preparation from natural and bio-sources [3, 14, 45].

For experimental research, L-histidine is considered due to its affordability over other amines, amino acids, ionic liquids, and synthetic materials. One of the significant advantageous impacts of L-histidine over other amino acids involves its biosynthesis in *E. coli* and *Salmonella typhi.* During the selectivity of Histidine in amino acids, L-histidine was preferred over D-histidine due to its natural biosynthesis occurrence and the side effects of the enantiomeric specificity and costs of D-histidine, L-histidine was preferred over D-histidine as shown in Fig. 3.

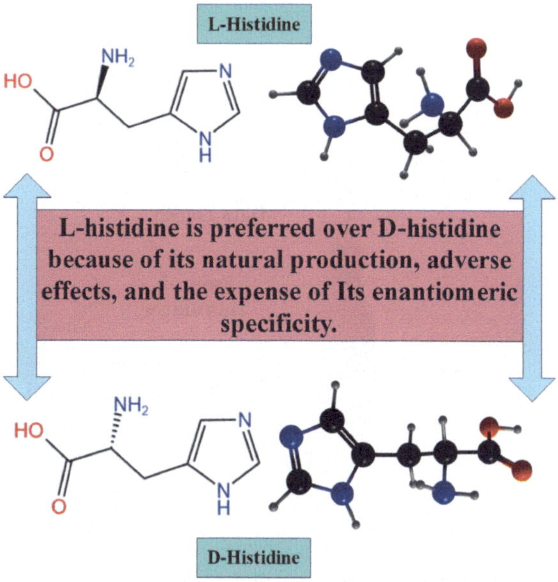

Fig. 3 L-histidine preference over D-histidine

2 Research Schematics and Setups

Schematics of the experimental setup for CO_2 uptake analysis consisted of L-histidine with 99% purity, reverse osmosis, membrane purified water, carbon dioxide cylinders, 100 ml stainless steel cylindrical reactor, data logger, thermocouple, pressure transducers, valves, regulator and indicator as shown in Fig. 4.

The procedure involved charging a 100 ml reactor with 80 ml absorbent material. At t = 0, the system is exposed to an initial pressure below 15 bar, and reduction is calculated in the data logger (every 30 s). Once the system reaches a steady state, stepwise injection is done. It is a three-injection setup in which stepwise injection is done thrice. Hence, the unabsorbed CO_2 was calculated to find the uptake capacity—the reactor setup involved 0.1 M L-histidine in water.

3 Method

For necessary calculations of the amounts of gases consumed, the following correlation is used in Eq. (1).

$$(\Delta n_{cons}) = V_g\left[\frac{P}{ZRT}\right]_0 - V_g\left[\frac{P}{ZRT}\right]_t \tag{1}$$

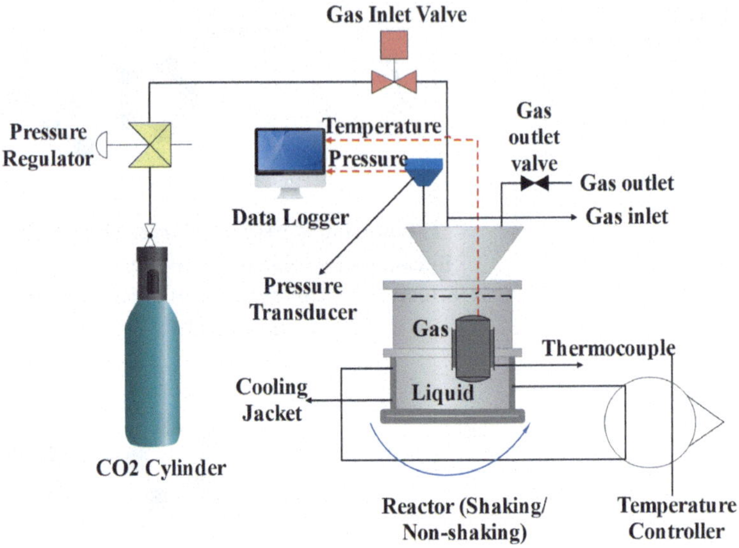

Fig. 4 Schematics of the experimental setup for carbon dioxide uptake analysis

Once the Pitzer correlations were simplified, a more straightforward equation is stated in Eq. (2) for the 100 ml reactor.

$$n = \frac{0.0075P}{z} \tag{2}$$

For the compressibility factor calculation, the Pitzer correlation is used primarily for gas sorption processes because such correlation defines the non-ideality of the gases in Eq. (3).

$$Z = 1 + \frac{BP}{RT} = 1 + \left(\frac{BP_c}{RT_c}\right)\frac{P_r}{T_r} \tag{3}$$

With further simplicities, Eq. (3) can be processed into Eq. 4 for a 100 ml reactor.

$$Z = 1 + 0.0016P \tag{4}$$

For all the variables shown in Eqs. (3) and (4), the following method is showcased in Eqs. (5) and (9).

$$\left(\frac{BP_c}{RT_c}\right) = Br = B^0 + \omega B^1 \tag{5}$$

$$B^0 = 0.083 - \frac{0.422}{T_r^{1.6}} \quad B^1 = 0.139 - \frac{0.172}{T_r^{4.2}} \tag{6}$$

$$Z = Z^0 + \omega Z^1 \tag{7}$$

$$Z^0 = 1 + B^0 \frac{P_r}{T_r} \quad \text{and} \quad Z^1 = B^1 \frac{P_r}{T_r} \tag{8}$$

$$Z = 1 + \frac{P_r}{T_r}\left(B^0 + \omega B^1\right) \tag{9}$$

where,

P → system pressure.

P_c→ critical pressure.

T_c→ critical temperature.

P_r→ reduced pressure.

T_r → reduced temperature.

ω→ acentric factor (0.224 for CO_2).

B → second virial coefficient.

R → ideal gas constant.

Z → compressibility factor.

B^0, Z^0, Z^1 → Empirical constants depend on T_r

4 Results and Discussion

Once the experimentation was completed, it was hypothesized that there was an uptake capacity of 49.67% in the first 30 min. Similarly, experimental analysis was done for 60, 90, and 120 min, as shown in Fig. 5.

The final uptake capacity after 2 h was 69.37%, whereas the total absorbed CO_2 after 2 h of the working reactor was 0.6116 g. Along with CO_2 uptake, the absorbed and unabsorbed CO_2 in the reactor is also portrayed in Fig. 6. The "point of equality" significantly showcases the point at which both the absorbed and unabsorbed CO_2 are exactly equal.

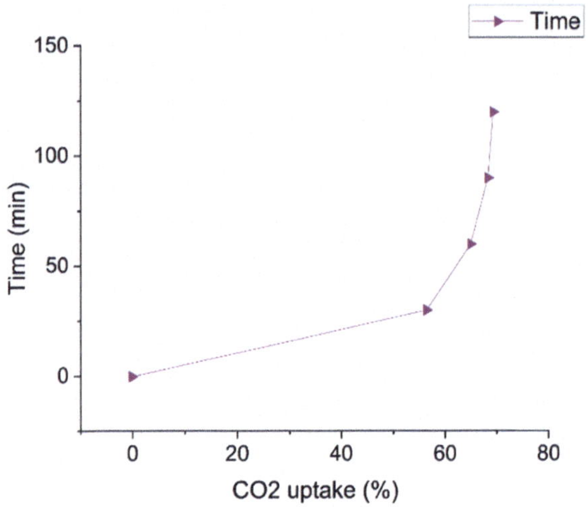

Fig. 5 The carbon dioxide uptake capacity of the system in a 100 ml reactor

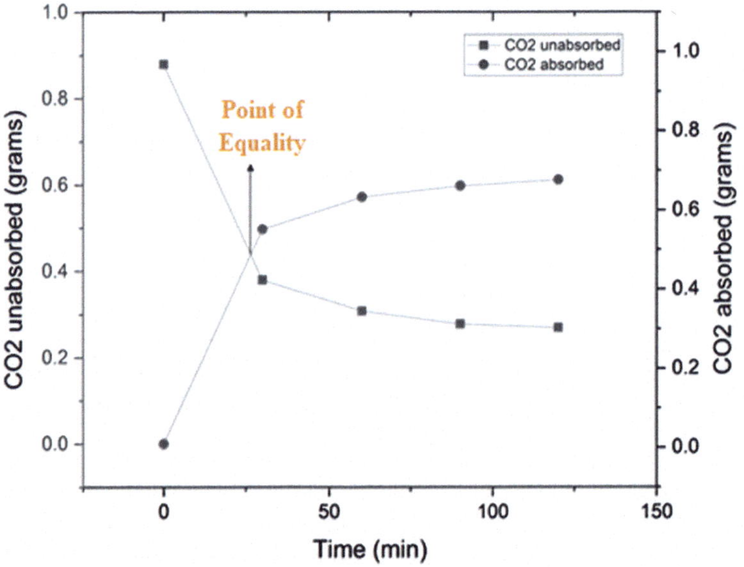

Fig. 6 Carbon dioxide absorbed and unabsorbed by the system (grams)

5 Industrial Scale-Up Perspective

From an industrial scale-up perspective, the current results were scaled up by a 100L reactor with L-histidine and water composition. The results in Table 2 show the total moles of CO_2 absorbed and unabsorbed in a 100L reactor. The absorbed and unabsorbed CO_2 was calculated by Pitzer correlation, taking 13 bar pressure and 100 L volume in the reactor (20 L gas volume); it mainly gives cross-estimation and approximation of the actual uptake capacity as well as absorbed CO_2 in 2 h for an industrial scale-up perspective. As the industries are run on ambient operating conditions (at room temperature (25–35 °C) and standard exterior atmospheric pressure), the setup was proportional to that as changes in temperature and pressure cause the calculations to be complex, making energy inefficient and expensive. Table 3 portrays the amount of CO_2 in grams for an industrial setup with a 100 L reactor.

In examining the industrial scaleup of a laboratory type setup there would be a substantial amount of elevation in the heat and mass transfer limitations which needs to be considered [43, 69]. These limitations predominantly arise due to the formation of stabilized carbamates, which can impede the mobility or diffusivity of CO_2 molecules within the aqueous system as the reaction proceeds, resulting in a decrease in the overall CO_2 absorption capacity of the system. Given the low concentration of amino acids in the aqueous system and its relatively low viscosity, the effect of mass and heat transfer limitations on the gas–liquid reaction is expected to be minimal. Therefore, to avoid overestimations we base our calculations on a

Table 3 Estimating the amount of CO_2 in grams for an industrial setup with the proposed 100 L reactor as a unit by considering ideal conditions

Time (min)	CO_2 unabsorbed (grams)	CO_2 absorbed (grams)
0	880	0
30	380	497
60	308	572
90	278.3	598
120	269.4	611.6

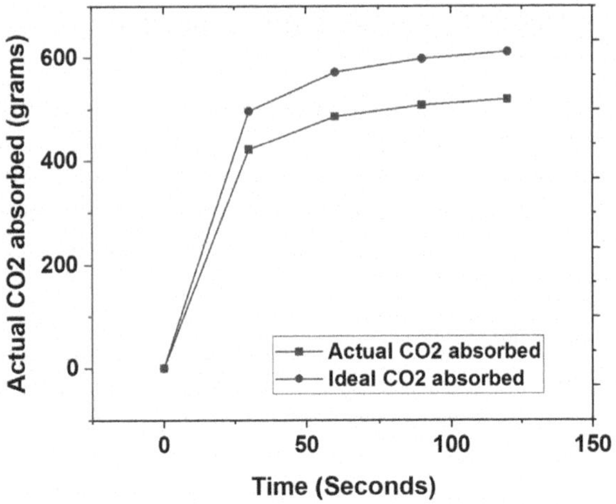

Fig. 7 Actual and Ideal CO_2 absorbed (in grams) in industrial scale up (100 L)

scenario where the CO_2 absorption into the system is decreased by roughly 10% from the ideal conditions due to these limitations and it is projected in Fig. 7.

6 Conclusion

Carbon dioxide storage facilities should be constructed with more effective units to bridge the gaps caused by ineffective policies, socioeconomic variables, the environment, and resource constraints. The main obstacles and gaps in the pursuit of carbon neutrality were the difficulty in adjusting net zero emission targets and the thermodynamic impossibility of carbon dioxide on its own, which can be resolved by reactivity with other reactants. Since energy accounts for the majority of carbon emissions (73.4%), it is imperative that energy transition towards zero carbon emissions—not merely towards electric vehicles or solar energy—be given top priority. In carbon capture, the Temperature Vacuum Swing Adsorption Model is the most promising technique due to its low installation cost, low risk, and low effect on impurities with maximum efficiency.

The Allam cycle, an oxy-combustion process, demonstrated the highest power generation efficiency compared to several CCPs for natural gas and bituminous coal. The chemical transformation of carbon dioxide was covered in great detail. However, several obstacles, such as high operating conditions, intricate reaction pathways, low product yield and conversion, and catalyst regeneration, should be considered in future studies. One of the significant innovations in CO_2 conversion is microalgae conversion, yet they face the same problems of high prices and low output. The annual carbon dioxide storage, estimated to have nearly 900 $MtCO_2$ of storage in 2030 but near zero $MtCO_2$/Year from 2020 to 2023, was the most shocking aspect of CCUS.

Due to the negligible carbon footprints of L-histidine, with no toxic gas emissions once dumped on land with atmospheric oxygen, it is considered one of the sustainable alternatives for novel carbon capture technologies. L-histidine's carbon dioxide uptake capacity is noted to be 69.37%, which is significant for industries with intermittent carbon dioxide emissions. Such uptake capacity makes it eligible for many scale-up opportunities, especially in sectors with more than 100L reactor capacities. A near-to-perfect absorbed and unabsorbed CO_2 was calculated, with nearly 75% of carbon dioxide (in grams) absorbed after two hours. A simplistic experimental schematic is presented with the least energy and materials. It is suitable for pilot scale-up as the next step for implementation in industries to target net neutrality and sustainable development goals.

Acknowledgements The authors equally acknowledge the IITM, ICT-IOC Bhubaneshwar, USCT, and GGSIPU for providing excellent research facilities and a supportive atmosphere. OSK acknowledges the CCUS Centre of Excellence IIT Madras, and HM acknowledges the Vice Chancellor of GGSIPU. The authors would also like to acknowledge MOKSH Research and Development for providing them with funding and the related facilities to ease this research. PK expresses gratitude to the Bhubaneshwar City Knowledge Innovation Cluster Foundation (BCKIC) and Kalinga Institute of Industrial Technology (KIIT) for honoring this work with the Best Poster Award as well as HM expresses gratitude for honoring this work with the best oral presentation award by INEEBA-2023 (Indo-South Korea- Thailand) international conference.

References

1. Al-Mamoori A, Krishnamurthy A, Rownaghi AA, Rezaei F (2017) Carbon capture and utilization update. Energ Technol. https://doi.org/10.1002/ente.201600747
2. Antonini C, Treyer K, Streb A, van der Spek M, Bauer C, Bauer C, Mazzotti M, Mazzotti M (2020) Hydrogen production from natural gas and biomethane with carbon capture and storage–a techno-environmental analysis. Sustain Energy Fuels. https://doi.org/10.1039/d0se00222d
3. Babu P, Ong HWN, Linga P (2016) A systematic kinetic study to evaluate the effect of tetrahydrofuran on the clathrate process for pre-combustion capture of carbon dioxide. Energy. https://doi.org/10.1016/j.energy.2015.11.009
4. Bajpai S, Shreyash N, Singh S, Memon AR, Sonker M, Tiwary SK, Biswas S (2022) Opportunities, challenges and the way ahead for carbon capture, utilization and sequestration (CCUS) by the hydrocarbon industry: towards a sustainable future. Energy Rep 8:15595–15616. https://doi.org/10.1016/j.egyr.2022.11.023
5. Belmabkhout Y, Serna-Guerrero R, Sayari A (2010) Amine-bearing mesoporous silica for CO2 removal from dry and humid air. Chem Eng Sci. https://doi.org/10.1016/j.ces.2010.02.044
6. Ben-Mansour R, Habib MA, Habib MA, Bamidele OE, Bamidele OE, Basha M, Basha M, Qasem NAA, Peedikakkal AMP, Laoui T, Ali M, Ali M, Ali M, Ali M, Ali M, Ali MdH (2016) Carbon capture by physical adsorption: Materials, experimental investigations and numerical modeling and simulations-a review. Appl Energy. https://doi.org/10.1016/j.apenergy.2015.10.011
7. Bhown AS, Freeman B (2011) Analysis and status of post-combustion carbon dioxide capture technologies. Environ Sci Technol. https://doi.org/10.1021/es104291d

8. Bobicki ER, Liu Q, Liu Q, Liu Q, Liu Q, Liu Q, Xu Z, Zeng H (2012) Carbon capture and storage using alkaline industrial wastes. Prog Energy Combust Sci. https://doi.org/10.1016/j.pecs.2011.11.002
9. Bui M, Adjiman CS, Stolten D, Bardow A, Anthony EJ, Boston A, Boston A, Brown S, Fennell PS, Fuss S, Galindo A, Hackett LA, Hallett JP, Herzog HJ, Jackson G, Jackson G, Kemper J, Krevor S, Maitland GC, Dowell NM (2018) Carbon capture and storage (CCS): the way forward. Energy Environ Sci. https://doi.org/10.1039/c7ee02342a
10. Chapman RG, Ostuni E, Takayama S, Holmlin RE, Yan L, Whitesides GM (2000) Surveying for surfaces that resist the adsorption of proteins. J Am Chem Soc 122(34):8303–8304. https://doi.org/10.1021/ja000774f
11. Chen S, Liu J, Zhang Q, Teng F, McLellan BC (2022) A critical review on deployment planning and risk analysis of carbon capture, utilization, and storage (CCUS) toward carbon neutrality. Renew Sustain Energy Rev 167:112537. https://doi.org/10.1016/j.rser.2022.112537
12. Chen X, Sun J, Sun J, Wang J, Cheng W (2012) Polystyrene-bound diethanolamine based ionic liquids for chemical fixation of CO2. Tetrahedron Lett. https://doi.org/10.1016/j.tetlet.2012.03.058
13. Chen XY, Vinh-Thang H, Ramirez AA, Ramirez AA, Rodrigue D, Kaliaguine S (2015) Membrane gas separation technologies for biogas upgrading. RSC Adv. https://doi.org/10.1039/c5ra00666j
14. Cormos C-C, Cormos C-C (2016) Oxy-combustion of coal, lignite and biomass: A techno-economic analysis for a large scale Carbon Capture and Storage (CCS) project in Romania. Fuel. https://doi.org/10.1016/j.fuel.2015.12.005
15. Cuéllar-Franca R, Azapagic A (2015) Carbon capture, storage and utilisation technologies: a critical analysis and comparison of their life cycle environmental impacts. J CO_2 Util. https://doi.org/10.1016/j.jcou.2014.12.001
16. Cui S, Cui S, Cheng W, Shen X, Fan M, Russell AG, Wu Z, Yi X (2011) Mesoporous amine-modified SiO2 aerogel: a potential CO2 sorbent. Energy Environ Sci. https://doi.org/10.1039/c0ee00442a
17. D'Alessandro DM, Smit B, Long JR (2010) Carbon dioxide capture: prospects for new materials. Angew Chem. https://doi.org/10.1002/anie.201000431
18. Ding H, Zhang Y, Dong Y, Wen C, Yang Y (2023) High-pressure supersonic carbon dioxide (CO2) separation benefiting carbon capture, utilisation and storage (CCUS) technology. Appl Energy 339:120975. https://doi.org/10.1016/j.apenergy.2023.120975
19. Ding M, Flaig RW, Jiang H-L, Yaghi OM (2019) Carbon capture and conversion using metal–organic frameworks and MOF-based materials. Chem Soc Rev. https://doi.org/10.1039/c8cs00829a
20. Fan X, Chao C, Chao C, Deng Y, Dewil R, Baeyens J, Baeyens J, Fan X (2015) Post-combustion carbon capture. Null. https://doi.org/10.1002/9781118991978.hces024
21. Fridahl M, Lehtveer M (2018) Bioenergy with carbon capture and storage (BECCS): global potential, investment preferences, and deployment barriers. Energy Res Soc Sci. https://doi.org/10.1016/j.erss.2018.03.019
22. Gan C, Liu Y, Liu Y, Tan X, Wang S, Zeng G, Zheng B, Li T, Jiang Z, Liu W, Liu W, Liu W (2015) Effect of porous zinc–biochar nanocomposites on Cr(VI) adsorption from aqueous solution. RSC Adv. https://doi.org/10.1039/c5ra04416b
23. He Q, O'Brien JW, Kitselman KA, Tompkins LE, Curtis GCT, Kerton FM (2014) Synthesis of cyclic carbonates from CO2 and epoxides using ionic liquids and related catalysts including choline chloride–metal halide mixtures. Catal Sci Technol. https://doi.org/10.1039/c3cy00998j
24. Heydari-Gorji A, Belmabkhout Y, Sayari A (2011) Polyethylenimine-impregnated mesoporous silica: effect of amine loading and surface alkyl chains on CO2 adsorption. Langmuir. https://doi.org/10.1021/la202972t
25. Hong WY (2022) A techno-economic review on carbon capture, utilisation and storage systems for achieving a net-zero CO2 emissions future. Carbon Capture Sci Technol 3:100044. https://doi.org/10.1016/j.ccst.2022.100044

26. Hospital-Benito D, Lemus J, Moya C, Santiago R, Palomar J (2020) Process analysis overview of ionic liquids on CO2 chemical capture. Chem Eng J. https://doi.org/10.1016/j.cej.2020.124509
27. Huertas JI, Gomez MD, Giraldo N, Garzón JP (2015) CO2 absorbing capacity of MEA. J Chem. https://doi.org/10.1155/2015/965015
28. Jiang K, Ashworth P (2021) The development of carbon capture utilization and storage (CCUS) research in China: a bibliometric perspective. Renew Sustain Energy Rev 138:110521. https://doi.org/10.1016/j.rser.2020.110521
29. Jiang K, Ashworth P, Zhang S, Liang X, Sun Y, Angus D (2020) China's carbon capture, utilization and storage (CCUS) policy: a critical review. Renew Sustain Energy Rev 119:109601. https://doi.org/10.1016/j.rser.2019.109601
30. Kong Y, Kong Y, Shen X, Shen X, Fan M, Yang M, Cui S, Cui S (2016) Dynamic capture of low-concentration CO2 on amine hybrid silsesquioxane aerogel. Chem Eng J. https://doi.org/10.1016/j.cej.2015.08.034
31. Kushwaha OS, Avadhani CV, Singh RP (2015) Preparation and characterization of self-photostabilizing UV-durable bionanocomposite membranes for outdoor applications. Carbohyd Polym 123:164–173. https://doi.org/10.1016/j.carbpol.2014.12.062
32. Leung DYC, Caramanna G, Maroto-Valer MM (2014) An overview of current status of carbon dioxide capture and storage technologies. Renew Sustain Energy Rev. https://doi.org/10.1016/j.rser.2014.07.093
33. Li X, Wang M, Wang C, Cheng C, Wang X (2014) Facile immobilization of Ag nanocluster on nanofibrous membrane for oil/water separation. ACS Appl Mater Interfaces 6(17):15272–15282. https://doi.org/10.1021/am503721k
34. Liang Z, Rongwong W, Liu H, Fu K, Gao H, Cao F, Zhang R, Sema T, Henni A, Sumon KZ, Nath D, Nath D, Gelowitz D, Gelowitz D, Gelowitz D, Srisang W, Saiwan C, Benamor A, Al-Marri MJ, Shi H, Supap T, Chan C, Zhou Q, Tontiwachwuthikul P (2015) Recent progress and new developments in post-combustion carbon-capture technology with amine based solvents. Int J Greenhouse Gas Control. https://doi.org/10.1016/j.ijggc.2015.06.017
35. Lin Q, Zhang X, Wang T, Zheng C, Gao X (2022) Technical perspective of carbon capture, utilization, and storage. Engineering 14:27–32. https://doi.org/10.1016/j.eng.2021.12.013
36. Lonkar SP, Kushwaha OS, Leuteritz A, Heinrich G, Singh RP (2012) Self photostabilizing UV-durable MWCNT/polymer nanocomposites. RSC Adv 2(32):12255. https://doi.org/10.1039/c2ra21583g
37. Markewitz P, Kuckshinrichs W, Leitner W, Linssen J, Zapp P, Bongartz R, Schreiber A, Müller T, Müller T (2012) Worldwide innovations in the development of carbon capture technologies and the utilization of CO2. Energy Environ Sci. https://doi.org/10.1039/c2ee03403d
38. Mason JA, Sumida K, Herm ZR, Herm ZR, Krishna R, Long JR (2011) Evaluating metal–organic frameworks for post-combustion carbon dioxide capture via temperature swing adsorption. Energy Environ Sci. https://doi.org/10.1039/c1ee01720a
39. Merkel TC, Lin H, Wei X, Baker RW (2010) Power plant post-combustion carbon dioxide capture: an opportunity for membranes. J Membr Sci. https://doi.org/10.1016/j.memsci.2009.10.041
40. Metz B, Davidson O, de Coninck HC, Loos M, Loos M, Meyer LH, Meyer L (2005) IPCC special report on carbon dioxide capture and storage. Null. null
41. Mittal H, Kushwaha OS (2024) Machine learning in commercialized coatings. In: Functional coatings. Wiley, pp 450–474. https://doi.org/10.1002/9781394207305.ch17
42. Mittal H, Verma S, Bansal A, Singh Kushwaha O (2024) Low-carbon hydrogen economy perspective and net zero-energy transition through proton exchange membrane electrolysis cells (PEMECs), anion exchange membranes (AEMs) and wind for green hydrogen generation. Qeios. https://doi.org/10.32388/9V7LLC
43. Oko E, Wang M, Ramshaw C (2019) Study of mass transfer correlations for rotating packed bed columns in the context of solvent-based carbon capture. Int J Greenhouse Gas Control 91:102831. https://doi.org/10.1016/j.ijggc.2019.102831

44. Olabi AG, Olabi AG, Obaideen K, Obaideen K, Elsaid K, Wilberforce T, Sayed ET, Maghrabie HM, Abdelkareem MA (2022) Assessment of the pre-combustion carbon capture contribution into sustainable development goals SDGs using novel indicators. Renew Sustain Energy Rev. https://doi.org/10.1016/j.rser.2021.111710
45. Olateju B, Kumar A, Kumar A, Kumar A, Kumar A, Kumar A (2013) Techno-economic assessment of hydrogen production from underground coal gasification (UCG) in Western Canada with carbon capture and sequestration (CCS) for upgrading bitumen from oil sands. Appl Energy. https://doi.org/10.1016/j.apenergy.2013.05.014
46. Ostuni E, Chapman RG, Holmlin RE, Takayama S, Whitesides GM (2001) A survey of structure−property relationships of surfaces that resist the adsorption of protein. Langmuir 17(18):5605–5620. https://doi.org/10.1021/la010384m
47. Padurean A, Cormos C-C, Cormos C-C, Agachi PS, Agachi PS, Agachi PS (2012) Pre-combustion carbon dioxide capture by gas-liquid absorption for integrated gasification combined cycle power plants. Int J Greenhouse Gas Control. https://doi.org/10.1016/j.ijggc.2011.12.007
48. Psarras P, Comello S, Bains P, Charoensawadpong P, Reichelstein S, Wilcox J (2017) Carbon capture and utilization in the industrial sector. Environ Sci Technol. https://doi.org/10.1021/acs.est.7b01723
49. Rabe M, Verdes D, Seeger S (2011) Understanding protein adsorption phenomena at solid surfaces. Adv Coll Interface Sci 162(1–2):87–106. https://doi.org/10.1016/j.cis.2010.12.007
50. Ramdin M, de Loos TW, Vlugt TJH (2012) State-of-the-art of CO2 capture with ionic liquids. Ind Eng Chem Res. https://doi.org/10.1021/ie3003705
51. Rashidi NA, Yusup S, Hameed BH (2013) Kinetic studies on carbon dioxide capture using lignocellulosic based activated carbon. Energy. https://doi.org/10.1016/j.energy.2013.08.050
52. Sanna A, Dri M, Hall MR, Maroto-Valer MM (2012) Waste materials for carbon capture and storage by mineralisation (CCSM)-a UK perspective. Appl Energy. https://doi.org/10.1016/j.apenergy.2012.06.049
53. Scholes CA, Smith KH, Kentish SE, Stevens GW, Stevens GW, Stevens GW (2010) CO2 capture from pre-combustion processes—strategies for membrane gas separation. Int J Greenhouse Gas Control. https://doi.org/10.1016/j.ijggc.2010.04.001
54. Sethia G, Sayari A (2015) Comprehensive study of ultra-microporous nitrogen-doped activated carbon for CO2 capture. Carbon. https://doi.org/10.1016/j.carbon.2015.05.017
55. Singh Kushwaha O, Avadhani CV, Singh Tomer N, Singh RP (2014) Accelerated degradation study of highly resistant polymer membranes for energy and environment applications. In: Advances in chemical science, vol 3, no 2. www.seipub.org/sepacs
56. Singh RP, Kushwaha OS (2013) Polymer solar cells: an overview. Macromol Symp 327(1):128–149. https://doi.org/10.1002/masy.201350516
57. Sinha A, Darunte LA, Jones CW, Realff MJ, Kawajiri Y (2017) Systems design and economic analysis of direct air capture of CO2 through temperature vacuum swing adsorption using MIL-101(Cr)-PEI-800 and mmen-Mg2(dobpdc) MOF adsorbents. Ind Eng Chem Res. https://doi.org/10.1021/acs.iecr.6b03887
58. Sircar S, Golden TC (2000) Purification of hydrogen by pressure swing adsorption. Sep Sci Technol. https://doi.org/10.1081/ss-100100183
59. Spigarelli BP, Kawatra SK (2013) Opportunities and challenges in carbon dioxide capture. J CO2 Util. https://doi.org/10.1016/j.jcou.2013.03.002
60. Stuckert NR, Yang RT (2011) CO2 capture from the atmosphere and simultaneous concentration using zeolites and amine-grafted SBA-15. Environ Sci Technol. https://doi.org/10.1021/es202647a
61. Sumida K, Rogow DL, Mason JA, McDonald TM, Bloch ED, Herm ZR, Bae T-H, Long JR (2012) Carbon dioxide capture in metal-organic frameworks. Chem Rev. https://doi.org/10.1021/cr2003272
62. Supasitmongkol S, Styring P (2010) High CO2 solubility in ionic liquids and a tetraalkylammonium-based poly(ionic liquid). Energy Environ Sci. https://doi.org/10.1039/c0ee00293c

63. Tcvetkov P, Cherepovitsyn A, Fedoseev S (2019) Public perception of carbon capture and storage: a state-of-the-art overview. Heliyon. https://doi.org/10.1016/j.heliyon.2019.e02845
64. Theo WL, Lim JS, Hashim H, Mustaffa AA, Ho WS (2016) Review of pre-combustion capture and ionic liquid in carbon capture and storage. Appl Energy. https://doi.org/10.1016/j.apenergy.2016.09.103
65. Thiruvenkatachari R, Su S, An H, Yu XX (2009) Post combustion CO2 capture by carbon fibre monolithic adsorbents. Prog Energy Combust Sci. https://doi.org/10.1016/j.pecs.2009.05.003
66. Praveenkumar V, Kushwaha OS, Nagaraaj P, Kumar R (2024) Green and sustainable carbon dioxide capture in aqueous l-valine: experimental studies on process efficiency and gas uptake for industrial applications. Ind Eng Chem Res 63(1):872–885. https://doi.org/10.1021/acs.iecr.3c03435
67. Veawab A, Tontiwachwuthikul P, Chakma A (1999) Corrosion behavior of carbon steel in the CO2 absorption process using aqueous amine solutions. Ind Eng Chem Res. https://doi.org/10.1021/ie9901630
68. Wall T (2007) Combustion processes for carbon capture. Null. https://doi.org/10.1016/j.proci.2006.08.123
69. Wurzbacher JA, Gebald C, Brunner S, Steinfeld A (2016) Heat and mass transfer of temperature–vacuum swing desorption for CO2 capture from air. Chem Eng J 283:1329–1338. https://doi.org/10.1016/j.cej.2015.08.035
70. Yan J, Wood DA, Zhang Z, Zhang Z (2019) Carbon capture, utilization and storage (CCUS). Appl Energy. https://doi.org/10.1016/j.apenergy.2018.11.019
71. Yan Y, Borhani TN, Subraveti SG, Pai KN, Prasad V, Rajendran A, Nkulikiyinka P, Asibor JO, Zhang Z, Shao D, Wang L, Zhang W, Yan Y, Ampomah W, You J, Wang M, Anthony EJ, Manovic V, Clough PT (2021) Harnessing the power of machine learning for carbon capture, utilisation, and storage (CCUS)–a state-of-the-art review. Energy Environ Sci 14(12):6122–6157. https://doi.org/10.1039/D1EE02395K
72. Yang Z, He L-N, Miao C-X, Chanfreau S (2010) Lewis basic ionic liquids-catalyzed conversion of carbon dioxide to cyclic carbonates. Adv Synth Catal. https://doi.org/10.1002/adsc.201000239
73. Yao X, Zhong P, Zhang X, Zhu L (2018) Business model design for the carbon capture utilization and storage (CCUS) project in China. Energy Policy 121:519–533. https://doi.org/10.1016/j.enpol.2018.06.019
74. Yokozeki A, Yokozeki A, Yokozeki A, Yokozeki A, Shiflett MB, Junk CP, Grieco LM, Foo T (2008) Physical and chemical absorptions of carbon dioxide in room-temperature ionic liquids. J Phys Chem B. https://doi.org/10.1021/jp805784u
75. Yu C-H, Huang C-H, Tan C-S (2012) A review of CO2 capture by absorption and adsorption. Aerosol Air Qual Res. https://doi.org/10.4209/aaqr.2012.05.0132
76. Zeleňák V, Badaničová M, Halamová D, Čejka J, Zukal A, Murafa N, Goerigk G (2008) Amine-modified ordered mesoporous silica : effect of pore size on carbon dioxide capture. Chem Eng J. https://doi.org/10.1016/j.cej.2008.07.025
77. Zhang B-T, Fan M, Bland AE, Bland AE, Bland AE (2011) CO2 Separation by a new solid K−Fe sorbent. Energy Fuels. https://doi.org/10.1021/ef200005x
78. Zhang Y, Zhang Y, Shao Z, Zhou W, Sunarso J, Liu S, Wang R (2013) Current status and development of membranes for co2/CH4 separation: a review. Int J Greenhouse Gas Control. https://doi.org/10.1016/j.ijggc.2012.10.009
79. Zhang Z, Lees EW, Habibzadeh F, Salvatore DA, Ren S, Simpson GL, Wheeler DG, Liu A, Berlinguette CP (2022) Porous metal electrodes enable efficient electrolysis of carbon capture solutions. Energy Environ Sci 15(2):705–713. https://doi.org/10.1039/D1EE02608A

Preparation of ABS/Mica Composites by Melt Mixing Method: A Critical Study of Mechanical and Thermal Properties

Supriya Senapati, Apsana Pradhan, Swarnalata Sahoo, and Aswini Kumar Mohapatra

Abstract This study explores the development of composites by homogeneously blending ABS (Acrylonitrile Butadiene Styrene) with natural calcined mica powder at varying mixing ratios. The melt mixing method in a twin-screw extruder was employed to prepare a mixture of virgin ABS granules and natural calcined mica powder, and the resulting compounded pellets were utilized for the fabrication of mechanical test specimens. Tensile, flexural, and impact tests were conducted according to relevant national and international standards, employing an injection molding machine. The investigation delves into the assessment of ABS/MICA composites across different ratios of ABS infusion, with a focus on mechanical, thermal, chemical, and physical characteristics. The interaction between ABS and MICA was confirmed through Fourier Transform Infrared (FTIR) analysis, providing insights into the molecular level bonding within the composite. Thermal stability of the composite was studied through TGA (Thermogravimetric) and DSC (Differential Scanning Calorimetry) analysis. Results revealed that the composite exhibited superior thermal stability compared to virgin ABS. These findings were further correlated with the mechanical properties of the composites, demonstrating a promising relationship between thermal stability and mechanical performance. The study indicates that increasing the loading of mica into ABS materials leads to a rise in brittleness in the resulting composites. Optimal properties were observed when the mica loading reached 15%, suggesting a balance between reinforcement and flexibility. In conclusion, the synergistic effects of ABS/MICA composites have been explored comprehensively, providing valuable insights into the interplay of materials and their resulting properties. This research contributes to the understanding of composite

A. Pradhan · S. Sahoo (✉)
Department of Plastic Engineering, Institute of Petrochemicals Technology (IPT), Central Institute of Petrochemicals Engineering and Technology (CIPET), Bhubaneswar, Odisha 751024, India
e-mail: sahoo.swarnalata@gmail.com

S. Senapati · A. K. Mohapatra (✉)
Department of Manufacturing Engineering and Technology, Institute of Petrochemicals Technology (IPT), Central Institute of Petrochemicals Engineering and Technology (CIPET), Bhubaneswar, Odisha 751024, India
e-mail: draswinikumarmohapatra@gmail.com

© The Author(s), under exclusive license to Springer Nature Singapore Pte Ltd. 2024
M. Suar et al. (eds.), *Advancements in Polymer Technology*, Springer Proceedings in Materials 57, https://doi.org/10.1007/978-981-97-6333-7_4

material design for applications that demand a combination of mechanical robustness, thermal stability, and chemical resistance.

Keywords ABS · MICA · Melt mixing · Composite · FTIR · DSC · TGA

1 Introduction

Polymer composites are polymer products with better mechanical, chemical, thermal, optical properties than individual polymers. Polymer composites are likely to contribute more in the plastic industries. Based on the interface attraction of fillers and polymers, the properties of the composite can be analyzed. In order to enhance the properties of the polymer in direction to mechanical, thermal and other physical properties, the inorganic mineral fillers are added. Based on the filler dimension and physical properties such as shape, size and distribution of filler, the improvement in mechanical properties of polymer composites strongly depends [1–5]. Acrylonitrile Butadiene Styrene (ABS) polymer is a thermoplastic amorphous opaque polymer of long chain of hydrocarbon $((C_8H_8)_x \cdot (C_4H_6)_y \cdot (C_3H_3N)_z)$ that exhibits a range of desirable characteristics, making it widely used in various industries. Being an amorphous material, it is not possible to find out true melting point but the particular glass transition temperature is approximately 105°Celsius (221°F). This polymer is made in the presence of polybutadiene ether by polymerising styrene and acrylonitrile. ABS is excellently stable structurally, which is why it is used in things like packaging, protective housings and camera housings. ABS is cheap, strong, stiff plastic that shows resistance towards external impacts. The unique properties of Acrylonitrile Butadiene Styrene (ABS) arise from the combination of its individual components, polybutadiene and poly (styrene-co-acrylonitrile), and the specific structure formed during polymerization. As the ABS mechanical properties vary with temperature, for the majority of applications it can be used between 20 and 80 degree Celsius. The effect of mica has been investigated in details on the properties of engineering thermoplastics. Mica has outstanding characterization properties less likely to be found in any other products. ABS/MICA composites, heralded for their cost-effectiveness and versatility, stand as a pioneering choice for a myriad of applications, notably excelling in the fabrication of automotive components, robust industrial hoses, and advanced packaging solutions, thereby revolutionizing sectors ranging from agriculture to aquaculture. We have studied the effect of mica with different concentrations on the properties of ABS thermoplastic such as ability, weather ability and porosity etc. The statistical analysis (abrasion F-test ANOVA) of the laboratory test samples with 10, 15, 20 and 25% mica in the composites show significant difference in thermal as well as mechanical properties [4–12].

In the realm of energy-efficient materials, the ABS/MICA composite emerges as a transformative force, showcasing immense promise as a structural thermal insulator that can revolutionize agriculture, aquaculture, and packaging industries. The paper delves into the distinctive material characteristics of ABS, an engineering thermoplastic renowned for its remarkable mechanical and thermal attributes, boasting a melting temperature reaching up to 245 °C. By incorporating particulate fillers like mica, known for its abundant availability and unique properties, the composite achieves an unparalleled synergy, enhancing both thermal and mechanical properties and paving the way for advancements in thermoplastic and thermoset applications [10–15].

ABS is commercialized in the year of 1940. It is a prominent thermoplastic used in many technical applications like refrigerators, machine prototyping, keyboard keys, etc. ABS is made tough with a rubbery component Butadiene. Styrene also gives ABS an excellent glossiness texture with excellent mechanical ability. However, ABS's overall mechanical characteristics are inferior to those of other popular engineering thermoplastics; particularly because of heat distortion temperature of standard ABS grades that are lower than 100 degree Celsius [9–11].

Most of the time, muscovite is found in silicic pegmatites that have granitic intrusions. Pegmatites without quartz are where phlogopite is found. Biotite comes from biotite-schists, which are metamorphic rocks. In some places, lepidolite and zinnwaldite are only found in granitic pegmatites. Micas are found as a whole in pegmatites and bands that run through mica schists. MICAs used for business are found in pegmatites as that are spread out by zone. Even though the pegmatite bodies aren't very big, the mica sheets are all different sizes. In the Bihar mica-belt, single books of muscovite that weigh half a ton have been found. Most pegmatites are made of plagioclase felspar and quartz, but they can also have orthoclase felspar and other crystals like tourmaline, garnet, apatite, and very rarely, beryl. They are often long, thin, lens-shaped, and, in some places, curved and winding. They have been placed along beneficial structures, such as the noses and limbs of folds, bedding, foliation, slip planes, and joints. They can be as small as a string and as big as 500 m long and 30 m wide. From a petro graphic point of view, there are different mappable units in a zoned or composite pegmatite with a quartz core, followed by a zone of quartz, felspar, and mica, and then a mica zone where the pegmatite vein meets the country rock. A very significant and sizable collection of minerals that are well suited for many uses includes mica. But because of its sophisticated features, it may be used in many different settings [11–18].

- Physically, mica is optically flat, colourless in thin sheets, readily divided into thin films along its cleavage, elastic, and incompressible.
- Chemical: It is a mixture of aluminium hydrous silicate with traces of lithium, magnesium, potassium, sodium fluorine, iron, magnesium, and many other elements. Water, acids (apart from hydrofluoric and concentrated sulphur), alkalis, ordinary solvents, bases, and oil have no effect on it; it is fully static and steady. The effects of the atmosphere have little impact on it.
- Electrically speaking, mica is the only material that combines uniform dielectric stability, capacitance stability, immense dielectric power, high Q factor and

decreased power loss, strong electrical resistance, and low temperature coefficient. It is praised for its resilience to corona discharge without resulting in any long-term damage.
- Thermally it can sustain temperatures up to 1000 °C (1832 °F) and is very fire resistant, incombustible, non-flammable, and infusible. But it also relies on the kind of Mica that is utilized. It is highly thermally stable, has decreased heat conductivity, and can withstand high temperatures without suffering any obvious consequences.
- Mechanically, mica is incredibly strong, elastic, and flexible. It also has a high tensile strength. It may be machined, die-punched, or hand cut and has a strong compressive force.

Engineering thermoplastic composites are mainly used in a wide variety of automotive parts such as sprockers, steers and gears etc. It is generally used for high strength industrial hoses in shock damping and tubing application.

The thermal conductivity of polymer composites is largely determined by the thermal conductivity networks in the matrix, and high thermal conductivity of polymer composites may be achieved by adding a high proportion of fillers. However, large filler content can negatively impact the processing and most of the mechanical characteristics of polymer composites. Recently researchers are focusing on filler modification and distribution in a polymer composite to improve the thermal conductivity of the polymer composite network. As a result, to create a highly thermal conductive or insulative polymer composite that meets all the requirements, a comprehensive approach is required. Evaluated ABS/MICA mix utilizing compatibilizers and examined the morphological and mechanical characteristics of ABS that had been grafted with maleic anhydride (MAH) [17–19].

The automobile sector is already beneficial by using this blend but in other sectors there are little or very limited use of ABS/MICA blends. So, we have targeted the fisheries sectors to provide them a cost effective, durable, heat insulating containers that can easily carry Juvenile fish for long distance transport facility. We conductedseveral tests such as impact, tensile, flexural, density, Fourier transform infrared spectroscopy, Differential scanning calorimetry, Thermogravimetric analysis, heat insulation where all test results indicate that this bend brings a new dimension and direction in the fisheries industry.

In the market already, there are some competitor materials available such as GFRP composite, but ey have a large drawback including low impact property because of the cracks generated and propagated in the impact load. And during unloading and uploading time when high impact strength use, that threaten the security of GFRP Container. While using ABS/MICA composites it has good impact strength and reduce the risk of cracking during unload and uploading time. Lower cost is another reason, ABS is cheap and almost half in price of its competitive materials that available in this sector and it is also easily available. ABS is lighter and easy to carry high strength to weight ratio. Moreover, due to blend with mica, it has good heat insulation properties.

Table 1 Materials and specification of ABS and MICA

Source and property	ABS	MICA
Function	Thermoplastic	filler
Grade	MIF-45 C-0001 NATURAL 8070	Calcinated mica powder mesh 100μm
Suppliers	Lucent polymers	Mica manufacturing Co. Pvt. Ltd. Mumbai, India
Density (ASTM D792)	1.05(g/cm^3)	2.05(g/cm^3)

2 Materials and Methods

2.1 Materials

The details of the materials as given below in Table 1.

2.2 Composite Preparation

To ensure moisture-free integrity, meticulous pre-drying of both the matrix and filler preceded compounding. ABS and MICA underwent a thorough 4-h drying process at 80 °C, with varying mica content (10, 15, 20, and 25%), and were adeptly blended using a co-rotating twin screw extruder (Model: MP19PC) operating at a consistent 60–80 rpm mixing speed (Table 2). Employing a precision temperature profile during extrusion, with Zone 1 at 180 °C, Zone 2 at 200 °C, Zone 3 at 210 °C, Zone 4 at 225 °C, and Die at 200 °C, the extrudites were expertly pelletized using a pelletizing machine with a controlled speed between 1.5–1.8 mm. Subsequently, the meticulously prepared compositions were transformed into specimens via an Injection moulding machine [ENDURA-90]. The details of the composition method is mentioned in Fig. 1.

Table 2 Different formulation ratios of the ABS and MICA composites

Sl. No	ABS wt%	MICA wt%	ABS/MICA
1	100	00	100:00
2	90	10	90:10
3	85	15	85:15
4	80	20	80:20
5	75	25	75:25

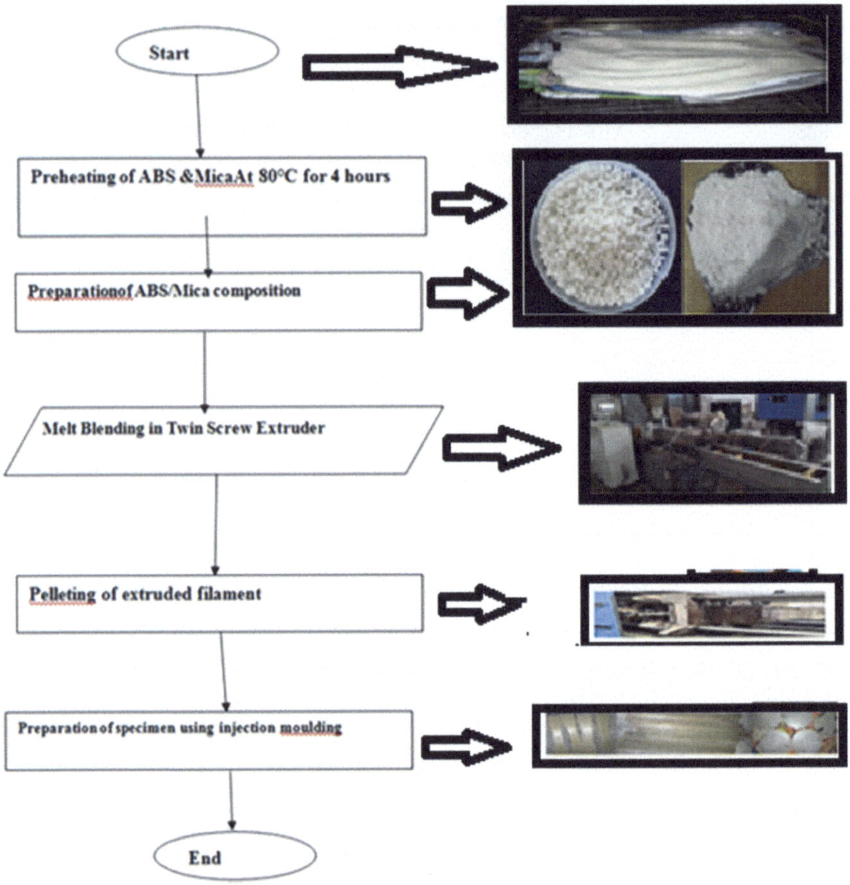

Fig. 1 Flow diagram of composites preparation

2.3 *Injection Moulding*

Following meticulous pre-drying in an air-circulated oven at 80 °C for 4 h, the granules of the extrudates underwent the transformation process through injection molding. This intricate procedure unfolded on a microprocessor-based ENDURA-90 injection molding machine, precisely configured to accommodate a master mold with cavities for crafting specimens designed for tensile strength, flexural, abrasion, and impact testing. Post-ejection from the mold, the specimens were systematically cooled at room temperature. Key processing parameters, including Zone 1 at 150 °C, Zone 2 at 225 °C, and Zone 3 at 245 °C, were rigorously controlled to ensure the production of specimens with consistent and optimal mechanical properties.

Table 3 Mechanical properties of different compositions of ABS with ABS-MICA composites

Sl.no	Different composition specimen	Tensile strength (MPa)	Tensile modulus (MPa)	Flexural strength (MPa)	Impact strength (J/m)
1	VIRGIN ABS	41.5 ± 1.3	4131.6 ± 14	66.74 ± 10	99.0696
2	90%ABS/10%MICA	30.28 ± 1.8	4232.5 ± 18	–	29.361
3	85%ABS/15%MICA	35.04 ± 1.4	4402.9 ± 25	73.25 ± 8	29.240
4	80%ABS/20%MICA	43.01 ± 2.5	4462.9 ± 26	69.62 ± 12	27.173
5	75%ABS/25%MICA	44.1 ± 2.2	4480.5 ± 34	69.14 ± 14	25.471

3 Results and Discussions

3.1 *Impact Test (Impact Strength, Make: Tinius Olsen ASTM D 256)*

Table 3 shows impact properties of virgin ABS, different wt.% of ABS/MICA composite. From Fig. 2 it is observed that there is a decrease in impact strength with the increase in MICA content (wt.%). It is clearly seen from Fig. 2, that the ratio 75:25 exhibits the lowest impact strength. This is due to the rigidness of Mica within the polymer matrix. Further, it is also observed that 90:10 shows a better impact property. Further, the addition of 15% MICA into ABS shows similar results of impact strength as 10% MICA into ABS [4].

3.2 *Tensile Test (Tensile Strength and Modulus, Make: Instron, ASTMD 638)*

Figures 3 and 4 reveal the dynamic relationship between tensile strength and Tensile Modulus in ABS and ABS/MICA composites across different weight percentages. Notably, Fig. 3 highlights a progressive enhancement in tensile strength with low tensile modulus corresponding to increased MICA loading, with the pinnacle achieved at a 25% MICA loading by employing tensile pull rate 50 mm/min. This uptick is attributed to the reinforcing prowess of MICA within the ABS polymer matrix, where the robust and rigid MICA filler adeptly facilitates stress transfer within the matrix. This reinforcing trend echoes in the case of both tensile strength and tensile modulus, exhibiting an incremental rise in accordance with the weight percentage of MICA loading compositions, as elucidated [4, 15]. It is also observed that at 10 and 15% loading of MICA within ABS decreases the tensile strength and modulus. The differences in tensile strength of the composite as compared to ABS are approx. 6–28%. This may be due to the improper dispersion of the filler within the matrix.

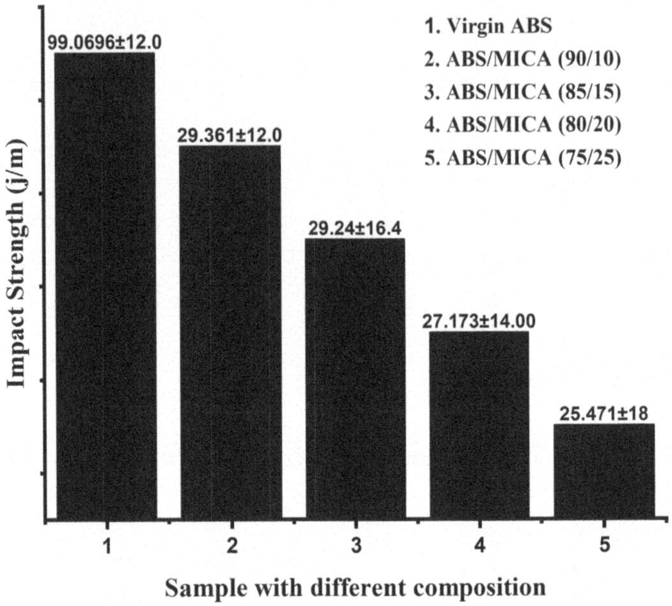

Fig. 2 Impact strength of Virgin ABS and ABS/MICA Composite

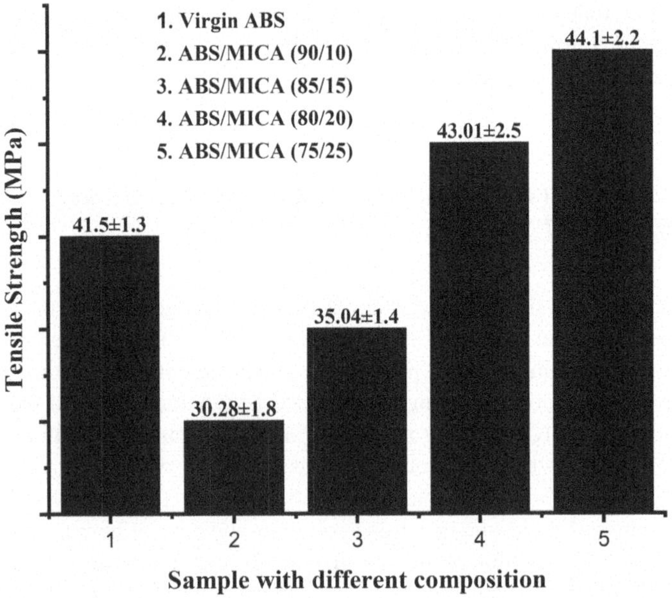

Fig. 3 Tensile strength of Virgin ABS, 90%ABS + 10%MICA, 85%ABS + 15%MICA

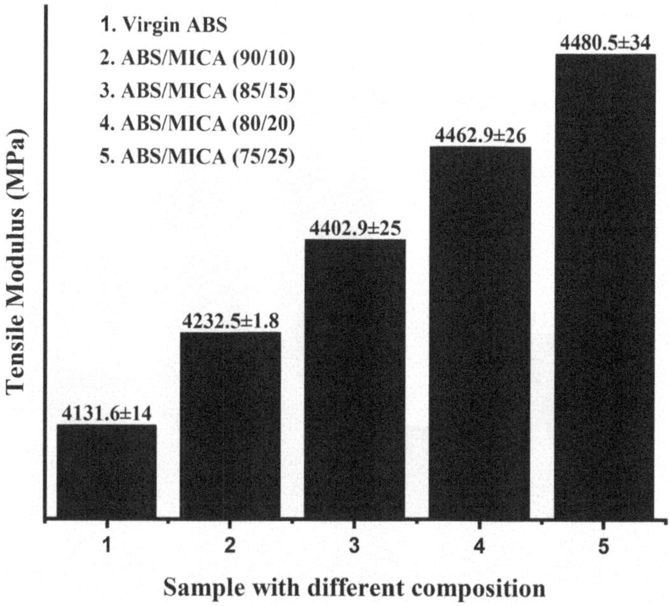

Fig. 4 Tensile modulus of ABS and ABS/ MICA different composition

3.3 *Flexural Test(Flexural Strength and Modulous, Make: Instron, ASTM D 790)*

Flexural characteristics also indicated a possible interaction between the matrix and the filler. In comparison to the virgin polymer the flexural strength is significantly improved with incorporation of MICA loading as we can see in Fig. 5 where the maximum flexural strength obtained in 85:15 wt. % ABS/MICA composition. This is due to the efficient stress transfer within the matrix and filler. As compared with the impact strength, the formulation of ABS/ MICA composite of 85:15 wt. % can be considered as the optimum ratio [1–4].

3.4 *Thermal Conductivity (Make: Unitherm USA, ASTM E 1530)*

The thermal conductivity (TC) test of ABS and ABS/MICA composite (85:15) has been studied. The thermal conductivity value of virgin ABS is found at 0.112, 0.134 w/mk at 30 and 60 °C respectively. Further, the results of the composite also comprise of TC measurement that has been carried out at 30 and 60 °C. It was observed that the set point:1 for 30 °C measurement having upper sensor is 43.49 °C, lower sensor is 21.54 °C and heat sink is 14.18 °C. That results when the supplied sample temperature

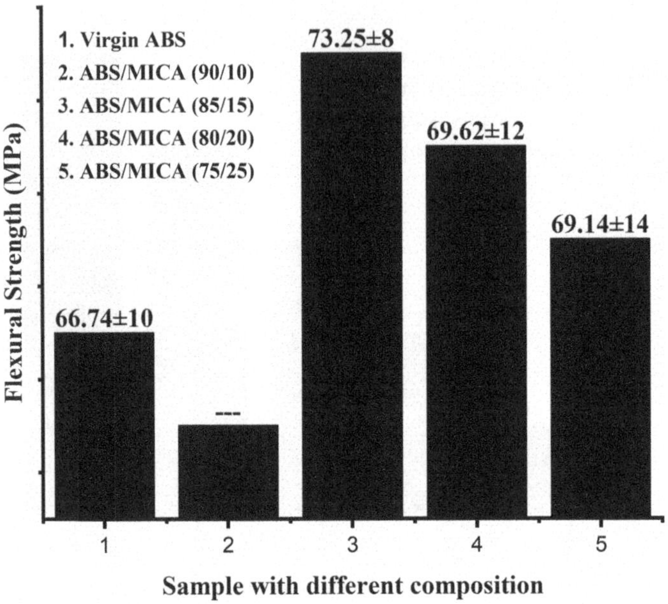

Fig. 5 Flexural strength of Virgin ABS, 90%ABS/ 10%MICA, 85%ABS/15%MICA

is 32.52 °C, the thermal conductivity is 0.148 w/mk and thermal resistivity is 1.99e-002m^{2}k/w. Similarly, for set point:2 for 60 °C measurement having upper sensor 73.18 °C, lower sensor 51.26 °C, and heat sink is 43.95 °C. That results when the supplied sample temperature is 62.22 °C, the thermal conductivity is 0.155 w/mk and thermal resistivity is 1.91e-002m^{2}k/w. Hence higher thermal conductivity is observed as compared with virgin ABS [20].

3.5 Differential Scanning Calorimetry (DSC) (Make: Perkin Elmer, ASTM D 3418)

Differential Scanning Calorimetry (DSC) indicates the ABS/MICA (85:15) composite phase changes over time and temperature in a meticulously controlled environment. This method provides a comprehensive blend of quantitative and qualitative insights into physical and chemical transformations, resolving the refinement of endothermic or exothermic processes and alterations in heat capacity. Within the realm of amorphous solids, ABS undergoes a transformative journey through the glass transition, a reversible shift from a brittle to a resilient state. Figure 6 elegantly illustrates the measured glass transition temperatures, unraveling the thermal dynamics not only of ABS but also of the ABS/MICA composites, adding a refined layer to the composite material's performance profile. The glass transition

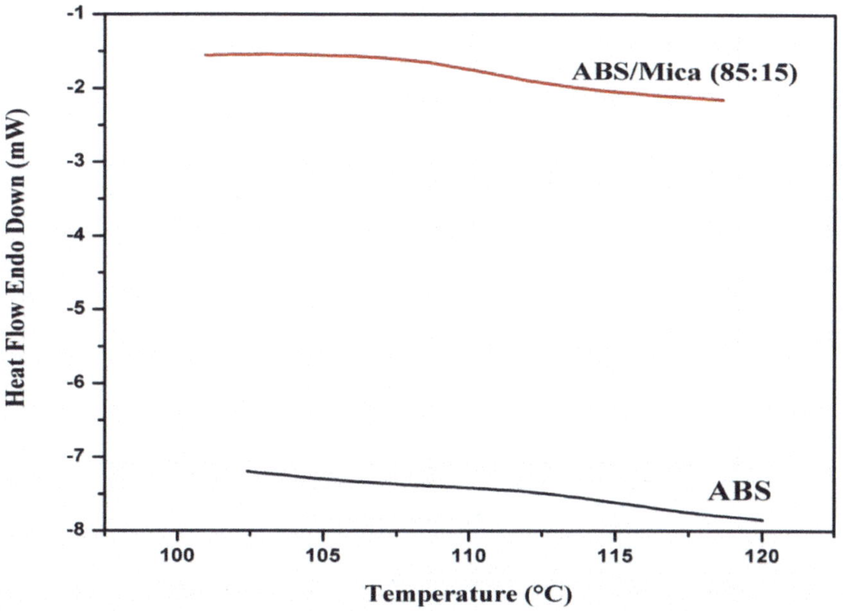

Fig. 6 Glass Transition thermogram of ABS and ABS/MICA

temperature of ABS is found at 108°C and after the addition of MICA 15 wt. % within the ABS, the glass transition temperature is shifted little towards the lower side and found at 105°C. Similarly, the melting temperature in both found around at 190 °C. Hence, no such differences have not been found in glass transition temperatures and melting temperatures in both virgin ABS and composite [18] (Fig. 7).

3.6 Thermogravimetric Analysis (TGA) (Make: Perkin Elmer, TGA 8000, ASTM E 1131)

The thermogravimetric analysis analysed the thermal intricacies of ABS and its MICA-filled composites under a nitrogen atmosphere, elegantly portrayed in the accompanying graph. Both thermograms exhibit a parallel behavior characterized by a dual-step degradation process. The primary degradation step, attributed to the breakdown of the polymer backbone, stands as a pivotal indicator of the system's thermal stability. A secondary, albeit minor, degradation step follows, associated with the disintegration of mica particle components. The correlation between the filler content and residue, showcased a proportional increase in residue as the mica content escalates, underscoring the significant role of mica in influencing the thermal characteristics of the composite system [8, 11, 18] (Table 4 and Fig. 8).

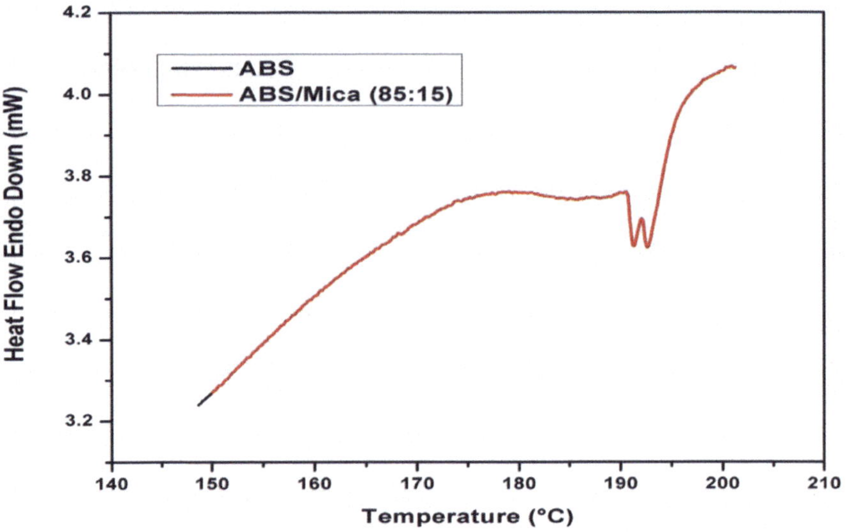

Fig. 7 Melting temperature of ABS and ABS/MICA

Table 4 Different steps of thermal degradation temperature of ABS and ABS/MICA

Sample code	$T_{initial}$ in °C	$T_{10\%}$ in °C	$T_{50\%}$ in °C	$T_{final\ \%}$ in °C	$T_{char\%}$ in °C
ABS	249.93	396.13	429.25	508.94	0.00
ABS/MICA	332.66	400.90	436.92	637.647	12.49

3.7 Fourier Transform Infrared (FTIR) Spectroscopy (Make: Perkin Elmer, ASTM E1252)

The FTIR analysis revealed the spectra for MICA, pure ABS and composites are given in Fig. 1. Mica being an inorganic material, it shows bands in the finger print region. ABS shows its characteristic peaks corresponding to the styrene group, peaks at 1445, 2240, 2845 and 2920 cm^{-1} which account for the different vibrations in the molecule [11, 18]. The intensity of the peak corresponding to mica is increased as the weight percentage of mica increases as evidenced from Fig. 1 (Table 5 and Fig. 9).

4 Conclusions

Incorporation of MICA (15 wt. %) within ABS results in improved properties especially higher in impact, flexural and thermal conductivity characteristics of the developed composites as compared with the other formulations of the composite. This composite shows overall improvement for further characterisation. From FTIR study

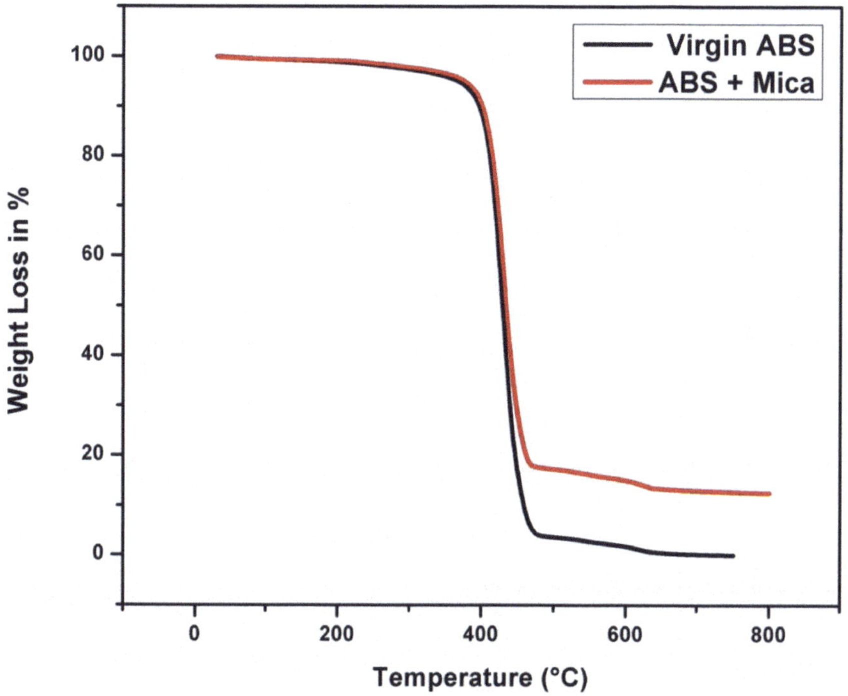

Fig. 8 TGA graph of ABS and ABS/MICA

Table 5 FTIR absorption peaks of ABS/MICA composites

Absorption (cm^{-1})	Group	Appearance
2920	N–H	Strong, broad
2845	N–H	Strong, broad
2240	C≡N	Weak
1445	C-H	Medium
1025	C-N	Medium
750	C-H	Medium
690	C = C	Strong

we get the involvement of different organic compounds within the matrix which leads to more stability as compared to virgin ABS, which is being improved by DSC and TGA analysis with improvement of melting temperature as well as melting glass transition and crystallisation temperature. The thermal conductivity values of ABS/MICA composites also show at 30 ºC is 0.148w/mk and at 60 ºC is 0.155w/mk. Thus, it may be concluded that the developed composite system facilitates in thermal insulation application, in electronic appliances and automotive industries.

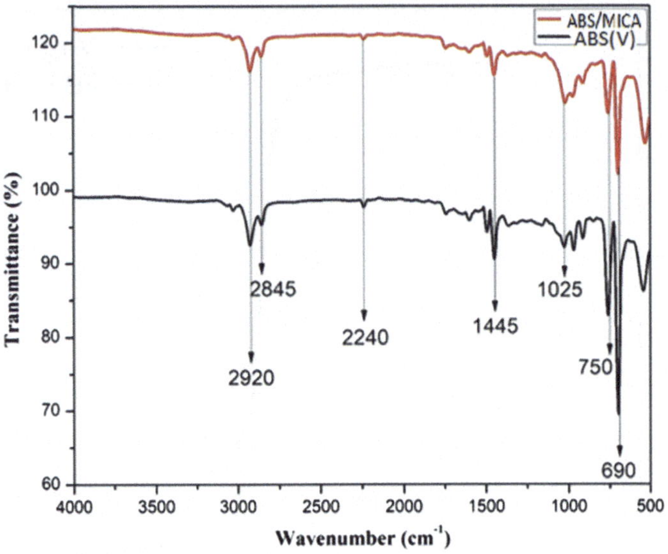

Fig. 9 FTIR graphs of ABS and ABS/MICA

References

1. Ahmed SA, Saleh SA, Abdel-Hameed SA, Fayad AM (2019) Catalytic, kinetic and thermodynamic properties of free and immobilized caseinase on mica glass-ceramics. Heliyon 5(5)
2. Jena SK, Dash N, Rath SS (2019) Effective utilization of lime mud for the recovery of potash from mica scraps. J Clean Prod 231:64–76
3. Bilotta MT, Abruzzese MP, Molfetta R, Scarno G, Fionda C, Zingoni A, Cippitelli M (2019) Activation of liver X receptor up-regulates the expression of the NKG2D ligands MICA and MICB in multiple myeloma through different molecular mechanisms. FASEB J 33(8):9489–9504
4. Asyadi F, Jawaid M, Hassan A, Wahit MU (2013) Mechanical properties of mica-filled polycarbonate/poly (acrylonitrile-butadiene-styrene) composites. Polym-Plast Technol Eng 52(7):727–736
5. Chang J-H, An YU, Cho D, Giannelis EP (2003) Poly (lactic acid) nanocomposites: comparison of their properties with montmorillonite and synthetic mica (II). Polymer 44(13):3715–3720
6. Souza DH, Andrade CT, Dias ML (2014) Isothermal crystallization kinetics of poly (lactic acid)/synthetic mica nanocomposites. J Appl Polym Sci 131(11)
7. Sreekanth MS, Joseph S, Mhaske ST, Mahanwar PA, Bambole VA (2011) Effects of mica and fly ash concentration on the properties of polyester thermoplastic elastomer composites. J Thermoplast Compos Mater 24(3):317–331
8. Nasif RA (2012) Study the effect of volume fraction of Mica with different particle size on the mechanical and electrical properties for unsaturated polyester composites. Eng Technol J 30(20):3573–3580
9. Abd AA (2016) Study the effect of mica as filler in natural rubber properties. J Univ Babylon 24(3):773–781
10. Gaval VR, Sahai RSN (2015) Effect of particle size and concentration of fly ash on properties of polytrimethylene terepthalate. In: International conference on chemical, mettallurgy and material science engineering (CMMSE 2015), pp 46–51. (Aug 2015)

11. Deshmukh SP, Rao AC, Gaval VR, Mahanwar PA (2011) Mica-filled PVC composites: effect of particle size, filler concentration, and surface treatment of the filler, on mechanical and electrical properties of the composites. J Thermoplast Compos Mater 24(5):583–599
12. Hemanth R, Sekar M, Suresha B (2014) Effects of fibers and fillers on mechanical properties of thermoplastic composites. Indian J Adv Chem Sci 2(January):28–35
13. Liang JZ, Yang QQ (2017) Effects of carbon fiber content and size on electric conductive properties of reinforced high density polyethylene composites. Compos B Eng 114:457–466
14. Bose S, Mahanwar PA (2004) Effect of flyash on the mechanical, thermal, dielectric, rheological and morphological properties of filled nylon 6. J Miner Mater Charact Eng 3(2):65–89
15. Almeida PDO, Gerardo CF, Leão AGD, França SCA, Santos SF, Bastos DC (2021) Sustainable composites based on recycled high-density polyethylene/mica. Mater Res 24
16. Sreekanth MS, Bambole VA, Mhaske ST, Mahanwar PA (2009) Effect of concentration of mica on properties of polyester thermoplastic elastomer composites. J Miner Mater Charact Eng 8(4):271–282
17. Juskey VP, Chaffey CE (1982) Rheology and tensile properties of polypropylene reinforced with glycerol-treated mica. Can J Chem Eng 60(3):334–341
18. Alghamdi M (2019) Acrylonitrile butadiene styrene (ABS)/mica composites: preparation and characterization. Minia J Eng Technol (MJET) 38(1):244–255
19. Lee GW, Park M, Kim J, Lee JI, Yoon HG (2006) Enhanced thermal conductivity of polymer composites filled with hybrid filler. Compos A Appl Sci Manuf 37(5):727–734
20. Tsukuda R, Sumimoto S, Ozawa T (1997) Thermal conductivity and heat capacity of ABS resin composites. J Appl Polym Sci 63(10):1279–1286

Synthesis and Characterization of Cellulose Acetate from Kombucha SCOBY Bacterial Cellulose Using Different Acetylating Agents

Laxmi Priya Swain, Muhil Raj Prabhakar, Bhaskar Das, and Paramasivan Balasubramanian

Abstract Kombucha bacterial cellulose (KBC) is an exopolysaccharide synthesized by bacteria in the process of fermenting kombucha tea. It has a distinct three-dimensional reticulated network made of cellulose nanofibers, which gives it robust mechanical qualities, a high water-holding capacity, and outstanding suspension stability. Cellulose acetate (CA) is a very important derivative generated via trans-esterification. As an economical ester of cellulose, CA is used in retractable films, fibers, filters, and tubes because of its good mechanical qualities, low density with larger surface area, biodegradability, and nontoxicity. This study focuses majorly on two aspects: (1) the production of high-value yields (cellulose acetate) by using low-value waste (tea processing waste) for the cultivation of bacterial cellulose to address the challenges associated with high production costs, plastic waste, and limitations of utilizing cellulose derivatives (2) Selection of the better acetylating agent among acetic acid and acetic anhydride for conversion of KBC. CA was synthesized via two methods: using acetic acid (CA1) and acetic anhydride (CA2) wherein using the first method, the % acetyl group of 11.82 was obtained, whereas a two-fold increase in the % acetyl group (23.65) was observed using the second method. Further, the characteristic peaks observed in FTIR analysis confirmed the acetylation of KBC. CA2 showed lower crystallinity and higher thermal stability than CA1. Thus, this study demonstrated the synthesis of CA using acetic anhydride as a better method from KBC for use as a sustainable alternative to existing plastic-based and plant-based cellulose acetates. The novelty of this research lies in its innovative approach and comparative study to produce cellulose acetate by utilizing low-value waste to

L. P. Swain
Department of Environmental Sciences, Sambalpur University, Burla, Odisha 768019, India

L. P. Swain · M. R. Prabhakar · P. Balasubramanian (✉)
Bioenergy and Environmental Lab, Department of Biotechnology and Medical Engineering, National Institute of Technology Rourkela, Rourkela, Odisha 769008, India
e-mail: biobala@nitrkl.ac.in

B. Das
KIIT-Technology Business Incubator, KIIT University, Bhubaneswar, Odisha 751024, India

© The Author(s), under exclusive license to Springer Nature Singapore Pte Ltd. 2024

M. Suar et al. (eds.), *Advancements in Polymer Technology*, Springer Proceedings in Materials 57, https://doi.org/10.1007/978-981-97-6333-7_5

achieve a high-value output like better yield, crystallinity, and thermal stability. It presents an environmentally friendly substitute for plastic-based cellulose acetates.

Keywords Cellulose acetate · Cellulose nanostructure · Kombucha bacterial cellulose · Biodegradable polymer

1 Introduction

Cellulose acetate is considered to be a significant derivative of cellulose, which is produced through a transesterification reaction. The chemical formula of cellulose acetate (CA) is $C_6H_7O_2(OH)_3$, which consists primarily of glucose molecules in its structure and has excellent properties, such as easy bonding with plasticizers, high flexibility, good chemical resistivity, and a variety of beneficial properties that make it suitable for several purposes [16]. As an economical ester of cellulose, it has many applications in industries such as the textile industry to the cigarette manufacturing process, and as a raw material for fibers, films for protecting Liquid Crystal Display (LCDs), and cosmetics. It is a viable option to replace the plastic film base of motion picture and sheet film [4]. CA is mainly used in cigarette filters, one of the most common kinds of waste that releases microplastic into the environment. Around 4.5 trillion used cigarette filters penetrate the environment annually, making them one of the most common types of plastic waste collected in urban and coastal cleanups over the past three decades [17]. This plastic pollution has an immediate and direct impact on all life forms. Consequently, research into an alternate raw material to synthesize CA has attracted the interest of scientists. The main ingredient for the synthesis of CA is cellulose, which is classified into five groups based on its source such as plant-based, wood-based, algae-based, bacteria-based, and tunicate-based [28]. The extraction of cellulose from wood, plants, algae, and tunicate is a very difficult task and also not economical. Plant-based cellulose contains a complex ribbon-like structure and some impurities such as hemicellulose, lignin, and pectin. Because of the presence of these impurities in plant cellulose, it is tough to degrade. It possesses a water-retentive capacity of 60% and an adequate degree of crystallinity and tensile strength. Bacterial cellulose (BC) is predominantly produced by bacterial strains such as *Acetobacter xylinum* and is mainly composed of cellulose, water, and extracellular polymeric substances (EPS). The cellulose content ranges from approximately 80% to 99%, making it a highly pure and structurally unique form of cellulose. Aside from that, it has better tensile strength, is hydrophilic, and can absorb up to 100% water [23]. The main component of it is cellulose, with documented cellulose content ranging from around 80% to 99%, indicating its great purity and distinctive structural characteristics.

Cellulose acetate is mostly synthesized from plant sources such as wood pulp by the use of acetic acid and acetic anhydride in the presence of sulphuric acid as a catalyst forming cellulose triacetate, which undergoes partial hydrolysis to achieve the desired degree of substitution [11]. Kouadri et al. [20] synthesized cellulose acetate from cellulose extracted from *Astragalus gombo* seeds with a mass yield of around 42%. Nigam et al. [24] extracted cellulose from *Parthenium hysterophorus* and used it to synthesize cellulose acetate for further preparation of bioplastic film for potential use in the food, medicine, and packaging industries. There are fewer studies utilizing native BC for synthesizing this cellulose derivative despite its superior properties compared to plant cellulose. Rahmayetty and Sulaiman [26] utilized wastewater from the Arenga starch industry (WWAS) as a fermentation medium for producing BC and optimized the conversion into cellulose acetate by varying the ratios of bacterial cellulose: acetic anhydride. BC can be grown in various mediums ranging from synthetic medium to cost-effective sources such as waste from agricultural industries, food processing industries, beverage industries, wine industry, and fruit and related byproducts, to name a few [7]. Fermented tea or kombucha is one such medium that can be utilized for growing cellulose-synthesizing bacteria to kombucha bacterial cellulose (KBC). Because of its potential to reduce environmental impact and increase long-term sustainability, KBC has attracted significant attention as a promising raw resource [21]. KBC exhibits exceptional traits comparable to those of other microbial cellulose, including a high degree of purity, a three-dimensional nanofiber network, a high level of crystallinity, robust mechanical properties, and a high degree of mouldability during the manufacturing process. Additionally, this material is also biocompatible [29] and can degrade faster as compared to petroleum-based and plant-based cellulose.

The synthesis of cellulose acetate has been accomplished using a number of different acetylation processes; however, to the best of the authors' knowledge, KBC has not been investigated as a potentially useful, cost-effective, and environmentally friendly source. In the existing literature, BC has been employed as a raw material through either commercial procurement or utilization as a composite with other polymers. Currently, there is a lack of comparative research regarding the synthesis methods of cellulose acetate from plant cellulose or bacterial cellulose. By using a comparative approach, this research aims to select a more efficient method for higher yield and degree of substitution for conversion of KBC into CA and also determine the usefulness of BC as a renewable source of raw material for CA production.

In this study, an attempt to produce high-value products by utilizing low-value waste, BC was formed using the waste from tea processing industries to overcome the high production cost, combat the plastic waste issue, and resolve the drawbacks of utilizing cellulose derived from other sources. BC was produced using a symbiotic culture of bacteria and yeast inoculated in a fermentation medium of waste tea leaves and sugar. Further, the produced KBC was modified via two treatment methods to synthesize a biodegradable cellulose acetate derivative. CA1 and CA2 were synthesized using acetic acid and acetic anhydride, respectively. The acetylated KBC, along with native KBC, were then subjected to structural and functional characterization using sophisticated instruments for confirming the acetylation and comparative analysis of the treatments.

2 Materials and Methods

2.1 *Synthesis of Cellulose*

A sweetened tea mixture was prepared by using 8 g/L of tea processing waste and 80 g/L of cane sugar (commercially available), which was subsequently subjected to fermentation at room temperature for 14 days after the addition of 10% inoculum. Following the duration of fermentation, a cellulose pellicle was found floating on the medium's surface, which was removed and rinsed with distilled water (DW). That pellicle is known as Symbiotic Culture of Yeast and Bacteria (SCOBY), a structure composed of cellulose that forms a three-dimensional mat. This mat is characterized by the coexistence of acetic acid bacteria and osmophilic yeast, which engage in a symbiotic interaction. The bacterium *Acetobacter xylinum* is responsible for the production of SCOBY cellulose, a secondary metabolite of fermentation, and has a significant role in this biological process. The population of yeast and bacterial cells exhibits an initial rise during the first 14 days of fermentation in both the broth and the SCOBY. Subsequently, a decline is seen in the cellulose mat, while the liquid broth consistently maintains a greater number of viable yeast and bacterial cells. Further, it was washed with 0.1N NaOH to remove the biotic components and non-cellulosic components. After washing once with NaOH, it was subjected to three subsequent washes with tap water until the pH reached 7.0. Then, the cellulose was dried using a hot air oven at 35 °C and ground into a fine powder using a commercial pulverizer until the entire amount of sample passed first through Sieve-7 (pore size of 4 mm approximately) and then Sieve-1 (pore size of 1 mm approximately). This pulverized cellulose was stored and used for further experiments.

2.2 *Preparation of Cellulose Acetate*

2.2.1 Using Acetic Acid

Cellulose acetate (CA1) was synthesized according to De Souza et al. [13], which is described as follows: Mercerization and bleaching were used as two pretreatments in the synthesis of CA1. In the process of mercerization, 5 g of pulverized KBC was added to an aqueous solution comprising 100 mL of acetic acid with a concentration of 92% (v/v) and 1.4% (v/v) sulfuric acid and stirred steadily at 80 °C for three hours. Then the samples were rinsed with DW, filtered, and dried for three hours at 50 °C. The mercerized samples were then bleached for one hour at 50 °C with continual agitation using an aqueous solution of hydrogen peroxide 34% (v/v) and sodium hydroxide 4% (w/v). After that, the samples were cleansed, filtered, and dehydrated at 50 °C for two hours.

2.2.2 Using Acetic Anhydride

Cellulose acetate (CA2) was synthesized according to Candido and Gonçalves [8] which is described as follows:

In the process of synthesizing CA2, 24 mL of glacial acetic acid was mixed with 5 g of KBC. The resultant mixture was agitated at 37 °C for 1 h. Then, a solution comprising 0.08 mL of sulfuric acid and 40 mL of glacial acetic acid was introduced to the resultant mixture and agitated for 45 min. Subsequently, the amalgamation was subjected to a cooling process until it reached a temperature of 18.3 ºC. The temperature was raised to 35 ºC, thereafter introducing 28 mL of acetic anhydride and 0.6 mL of sulfuric acid to that mixture. The resulting mixture was then stirred for a period of 1.5 h. Subsequently, a mixture comprising 10 mL of water and 20 mL of acetic acid was introduced incrementally into the reaction vessel and subjected to stirring for 1 h. Subsequently, the acquired specimen underwent a rinsing process utilizing distilled water until a neutral pH of 7.0 was reached. Then, the sample was filtered and dried at 40 ºC for 10 h.

3 Characterization

3.1 Degree of Substitution

The average amount of acetyl groups that replace hydroxyl groups in the glucosidic units is known as the "degree of substitution" (DS). The determination of the DS of the sample was conducted to confirm it as cellulose acetate. The identification of DS was carried out utilizing a saponification reaction. Specifically, 0.1 g of CA1 and CA2 were added separately to a mixture of ethanol and 0.25 M NaOH 5 mL each. The mixture was allowed to remain standing for a duration of 24 h. Subsequently, a volume of 10 mL of 0.25 M hydrochloric acid was added to the mix and allowed to remain undisturbed for 30 min. Then, the resultant mixture was subjected to titration employing a standardized solution of 0.25 M NaOH, and phenolphthalein was utilized as an indicator. The experiment was replicated three times. The calculation of the proportion of acetyl groups was performed utilizing the subsequent Eq. (1) as reported by Rodrigues Filho et al. [27]

$$\%AG = \frac{[\{(V_{bi} + V_{bt}) \times \mu_b - (V_a \times \mu_a)\} \times 43 \times 100]}{mca} \tag{1}$$

where:

$\%AG$ is the Percentage of acetyl group, V_{bi} is the volume of NaOH added to the system, V_{bt} is the volume of NaOH spent in titration, μ_b is the concentration of NaOH, V_a is the volume of HCl added to the system, μ_a is the concentration of HCl, molar weight of the acetyl group is 43, *mca* is the weight of the cellulose acetate sample.

The DS was calculated using the following Eq. (2) [15]

$$DS = \left(\frac{162 \times \%AG}{4305}\right) - (43 \times \%AG) \quad (2)$$

3.2 Fourier Transform Infrared Spectroscopy (FTIR)

An Attenuated total reflectance- FTIR (MN_{344}, Alpha ATR-FTIR, Bruker) was used to analyze FTIR spectra of the KBC and CA samples, with a wavenumber scan ranging from 400 cm^{-1} to 4000 cm^{-1} and resolution of 4.0 cm^{-1}.

3.3 Scanning Electron Microscopy/Energy Dispersive Spectroscopy (SEM–EDS)

The FLEX SEM 1000 was used to analyze the surface image. After two hours of drying at 50 ºC, the sample was treated to a gold coating using scattering equipment (SCANCOAT, PIRANI 501) operating at a pressure of 0.3 mbar and a voltage of 20 kV 35 s.

The elements present in the CA were analyzed by EDAX, AMETEK® to identify the elements present in the sample.

3.4 X-ray Diffraction (XRD)

The X-ray diffraction patterns were obtained with an Ultima IV X-Ray Diffractometer manufactured by Rigaku, Japan. The samples were analyzed at 2θ range of 5^0 to 100^0 and a scan rate of 5° per minute.

3.5 Differential Scanning Calorimeter (DSC)

The amount of heat flow in KBC, CA1, and CA2 was analyzed by using the NETZSCHDSC200F3 instrument in a range from room temperature to 250 °C and 160 °C, respectively. The DSC curves were obtained using a sealed aluminium pan under a nitrogen atmosphere (50 mL min^{-1}) with a heating and cooling rate of 10 °C min^{-1}.

4 Results and Discussion

As KBC is synthesized by bacteria as a byproduct of fermentation, they are reported to be pure unbranched α- cellulose without the presence of any other polymers like lignin and hemicellulose [6, 21]. After a fermentation period of 14 days, the KBC with a yield of 47–68 g/L was obtained. It was harvested and washed with 0.1N NaOH, followed by distilled water, till a neutral pH was reached. Subsequently, they were dried and pulverized for synthesizing cellulose acetate. 5 g of powdered KBC was taken for the synthesis of CA1 and CA2. Finally, 3.4 $\pm$ 0.36 g of CA1 was obtained while using acetic acid as an acetylating agent, whereas 4.1 $\pm$ 0.11 g of CA2 was obtained in the second method using acetic anhydride.

Acetic acid is a weaker acetylating agent than acetic anhydride, takes longer reaction periods and higher temperatures to acetylate cellulose. Acetic acid interacts with cellulose predominantly by direct esterification, where acetyl groups are transferred to hydroxyl (-OH) groups of cellulose molecules. Therefore, cellulose acetate synthesis with acetic acid requires extended reaction durations and high temperatures to enhance esterification and assure acetyl group insertion into the cellulose backbone. Whereas, acetic anhydride is a stronger acetylating agent and interacts quickly with cellulose at milder circumstances. Instead of direct esterification, acetic anhydride acetylation often involves cyclic interactions with cellulose hydroxyl groups. Compared to acetic acid, this method acetylates cellulose quicker and more efficiently, frequently resulting in more acetylation and better product characteristics.

The two acetylating agents in this study have different chemical properties and reactivity mechanisms, so we used two different processes for them to optimize the cellulose acetate synthesis process for each agent and maximize product quality and utility. The mechanism behind the conversion of cellulose to cellulose acetate using acetic acid involves the activation of cellulose as the first step, in which the cellulose is subjected to react with acetic acid and sulfuric acid (acid catalyst). The cellulose fibers in the BC undergo swelling when immersed in the acetic acid-sulfuric acid solution. This occurs due to the breakdown of intermolecular hydrogen bonds among cellulose chains. Then, the mixture is stirred steadily at 80 °C for three hours. The elevated temperature aids in accelerating the mercerization process, leading to increased accessibility of cellulose chains. Following activation, acetic acid is added to the system, a process known as esterification. When exposed to acetic acid, it reacts with hydroxyl groups, causing the replacement of acetyl groups for the hydroxyl ones. In esterification, the hydroxyl group functions as the nucleophile attacking the acetic acid, forming a covalent bond and expelling water in the process. Then, the final product (CA1) was bleached by using sodium hydroxide and hydrogen peroxide (oxidizing agent) to get rid of any impurities and acids. Similar kinds of observations were earlier reported for bacterial cellulose from *Gluconacetobacter xylinus* [4]. At the same time, the interaction between hydroxyl groups and acetic anhydride occurs when the acyl carbon center of the acetic anhydride molecule is targeted by a lone pair from the alcoholic hydroxyl group, resulting in the formation of an ester. The monomeric composition of cellulose acetate (CA) is comprised of glucose molecules that are devoid of water and connected through ether bonds, specifically β-1, 4 glycoside linkages. The anhydrous glucose is modified at positions 2, 3, and 6 by substituting acetyl groups for the hydroxyl groups [1].

The synthesized CA yield is calculated from the ratio of the weight of CA in grams to the weight of BC in grams. Equation (3) provides the calculation for the percentage of CA yield as a result [26].

$$Yield(\%) = \frac{weight\ of\ CA(g)}{weight\ of\ BC(g)} \times 100 \tag{3}$$

The yield (%) of CA synthesized from acetic acid and acetic anhydride using the above equation was 68 ± 7.2% and 82.66 ± 2.3%, respectively. Rahmayetty and Sulaiman [26] reported that a higher acetic anhydride concentration leads to better cellulose acetylation and yield in the case of bacterial cellulose, where *Acetobacter xylinum* was used for the production of BC. So, further study can be focused on the increase in the concentration of acetic anhydride to get a better yield and increase the acetyl content.

4.1 Degree of Substitution (DS)

With the uniform acetylation of CA1 and CA2, it was observed that the % of acetyl groups of CA2 (23.65) with a degree of substitution of 1.67 was obtained using acetic anhydride and the % of acetyl groups of CA1 (11.82) with a degree of substitution of 0.5 was obtained using only acetic acid. Das et al. [12] synthesized CA from rice husk and reported a maximum degree of substitution of 1.08 at 80 °C. However, Cheng et al. [11] reported a higher degree of substitution (approximately 2.5) and weight yield of CA synthesized from cotton byproducts by utilizing Iodine as a catalyst. Additionally, it was mentioned that the crystalline structure of cellulose, which tends to resist chemical reactions, was responsible for the decreased amount of acetylation. Since a large portion of the cellulosic fraction is probably crystalline and chemically resistant, the percentage of cellulose that can be acetylated is anticipated to be far lower than 30%. [4] synthesized CA through homogenous acetylation of commercial bacterial cellulose and further reported that the acetylation time is a potential tool for regulating the degree of substitution (DS) values, and products with lower DS values have more biodegradability potential. Thus, future work can focus on the acetylation time and utilization of non-acid-based catalysts such as iodine for a better degree of substitution and yield of BC acetylation.

4.2 Fourier Transform Infrared Spectroscopy (FTIR)

The FTIR of kombucha bacterial cellulose (Fig. 1) shows a sharp peak at 1040 cm^{-1}, which represents the presence of C-O functional group (stretching of primary alcohol), peak at 1610 cm^{-1} assigned to H–O–H bending and 3350 cm^{-1} represents O–H functional group [4]. The FTIR of CA1 shows a sharp peak at 1040 cm^{-1}, which represents the presence of the C–O functional group of polysaccharide group, a peak at 1237 cm^{-1} represents symmetrical C–O stretching, which indicates a functional group of acetyl, 1390 cm^{-1} represents CH deformation vibration; O–H bending vibrations in phenol functional group and 3460 represents O–H stretching (hydrogen bonded) functional group [19]. Similarly, FTIR of CA2 showed sharper and higher peak intensities at 1748 cm^{-1}, 1386 cm^{-1}, and 1249 cm^{-1} representing ester carbonyl, CH vibration, and CO stretching of ester, respectively, which are all characteristic of acetyl groups. The difference in peak intensities between CA1 and CA2 described the effectiveness of the acetylating agents used as well as the degree of acetylation. In particular, compared to CA1, the sharper and higher peak intensities for CA2 at characteristic wavenumbers linked to acetyl groups suggest a greater degree of acetylation. Thus, it can be concluded from the FTIR analysis that acetic anhydride was a better acetylating agent as compared to acetic acid. Similar results were also reported for the synthesis of CA from textile waste by Homem and Amorim [18], and from paper waste by Beraich et al. [5]. Table 1 represents the wavenumbers of the characteristic peaks and their corresponding functional groups for KBC, CA1, and CA2.

4.3 Scanning Electron Microscopy (SEM)

The surface morphological changes occurring after acetate treatment as compared to the untreated cellulose are shown in Fig. 2. Acetylated KBC (Fig. 2b and c) exhibit a more compact structure and give a sponge-like texture due to swelling by the addition of acetyl groups [4]. In the second step, the process of acetylation is observed in crystallites as a mechanism of erosion. This erosion begins on the outer surface of the crystallites and progresses toward their center [4]. In both the CA samples, the magnified observation reveals the presence of small dots, potentially suggesting the formation of nanoscale aggregates composed of individual cellulose microfibrils. A similar result is also reported by Amaral et al. [2] for cellulose derived from babassu coconut shells. Acetylation reagents could more easily enter the core of cellulose particles after their crystalline structure was destabilized and the amorphous area in cellulose was expanded. The acetate is predominantly adhered to the cellulose surface (Fig. 2b and c), which exhibited a rough texture, suggesting that the acetylation reaction primarily took place on the surface of the cellulose particles [9].

The EDS of kombucha bacterial cellulose (Table 2) represents the presence of 85% of carbon and 15% of oxygen elements, whereas the EDS of CA1 represents

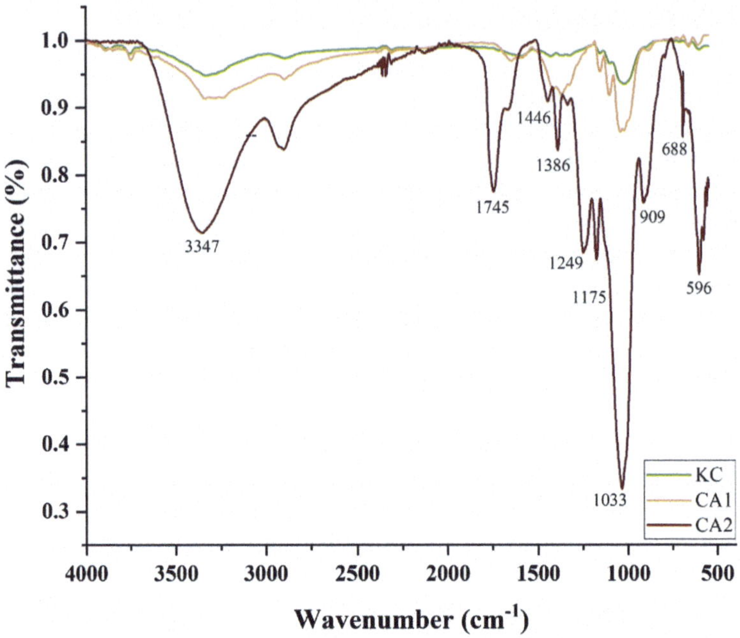

Fig. 1 FTIR analysis of Kombucha bacterial cellulose (KBC) and cellulose acetate (CA1 and CA2) samples

Table 1 Characteristic peaks obtained in native cellulose and acetylated cellulose samples and their corresponding functional group

Wave number (cm^{-1})			Functional group	References
KBC	CA1	CA2		
–	3460	–	O–H stretching (Hydroxyl group)	[22]
3350	–	–	O–H stretching (Carboxylic acid)	[14]
–	2800	–	C–H stretching (Aldehyde)	[14]
–	–	1748	Stretching of ester carbonyl	[22]
1610	–	–	C = C stretching (Aromatics)	[10]
–	1390	1386	C–H deformation vibration	[4]
–	1237	1249	C–O stretching (Acetyl group)	[4]
1040	1040	1040	Symmetrical C–O stretching	[4]

the presence of 50% of oxygen, 35% of carbon, and 15% of sodium elements and CA2 represents the presence of 53.5% of carbon, 45.9% of oxygen and 0.6% of sodium. In CA1 and CA2, sodium occurs due to the pretreatment of the KBC with sodium hydroxide. In CA1, % of sodium is more than in CA2 due to the action of the sodium hydroxide with the cellulose after mercerization. Amarasekara et al. [3] also reported the presence of sodium elements on the surface of the cellulose sample

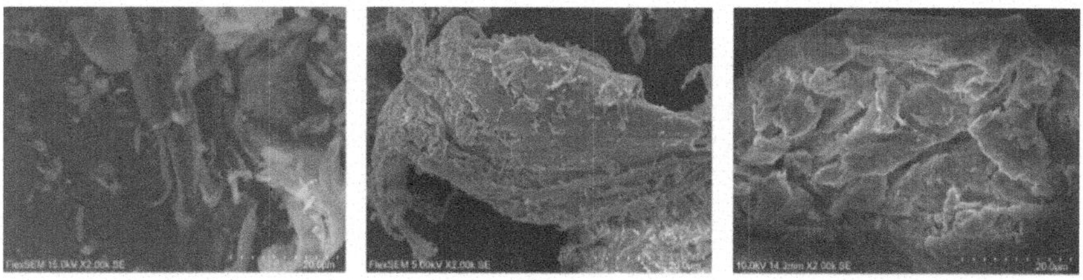

(a) (b) (c)

Fig. 2 SEM analysis of **a** KBC, **b** CA1, and **c** CA2 under 2000 × magnification

Table 2 EDS analysis showing the changes in the elemental composition of KBC, CA1, and CA2

Sample	Element	Weight %	Atomic %
KBC	C	85.0	88.3
	O	15.0	11.7
CA1	C	35.0	43.6
	O	50.0	46.7
	Na	15.0	9.7
CA2	C	53.5	60.6
	O	45.9	39.0
	Na	0.6	0.4

due to sodium hydroxide washing of cellulose. It further shows that the % of carbon decreases after acetylation.

4.4 X-ray Diffraction (XRD)

The XRD patterns of kombucha bacterial cellulose and cellulose acetates are represented in Fig. 3. Two main peaks were observed at 14.8° and 22.9°, which can be assigned to cellulose 1α and 1β phases as reported by Oliver-Ortega et al. [25] for *Acetobacters* and *Glucanobacters* strains. The primary feature of semi-crystalline acetylated derivative cellulose is said to be the major peak, which is situated at about 8°. The location of this peak shows that when the cellulose was acetylated, a disorder was created [4]. Acetylated cellulose is less crystalline than original cellulose because the substitution of bigger acetyl groups for smaller hydroxyl groups disrupts the hydrogen bonds within cellulose molecules [12].

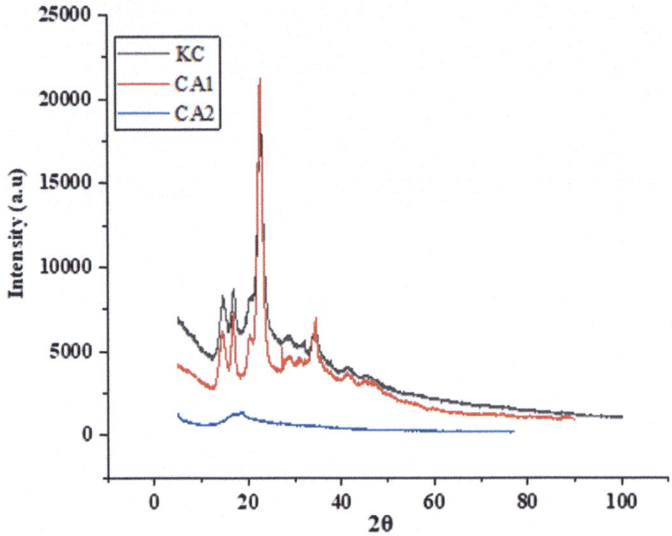

Fig. 3 XRD analysis of Kombucha bacterial cellulose (KBC) and cellulose acetate (CA)

4.5 Differential Scanning Calorimeter (DSC)

The heat-withstanding property, melting temperature, and glass transition temperature of KBC before and after treatment of acetate were observed using DSC. The DSC result (Fig. 4) of KBC shows a glass transition temperature (T_g) at 25 °C and an endothermic peak at 250 °C as its melting temperature (T_m). Rodrigues Filho et al. [27] reported a similar result when they synthesized CA from recycled newspaper. They observed endothermic behavior at approximately 265 °C, which is likely attributed to the material undergoing a phase transition, specifically the process of melting. CA1 is showing T_g at 30 °C and the endothermic peak at 160 °C (T_m). This peak is linked with the degradation of partially hydrolyzed cellulose. At the outset of the process, acid treatment hydrolyzes cellulose chain segments in ordered and dense crystallite areas, degrading the molecule and reducing its thermal stability. Similarly, CA2 is showing T_g at 27 °C and an endothermic peak of T_m at 220 °C. This signifies the existence of a crystallizable fraction of the material and a shift in the DS value, which may occur under certain environmental circumstances. When cellulose derivatives undergo DS changes, their thermal characteristics become quite sensitive. Similar results were also reported by Barud et al. [4], in which the endothermic peak of the CA sample showed in the range of 160–300 °C. Thus, it can be concluded from the DSC graph that CA2 shows good thermal stability and a high degree of substitution as compared to CA1.

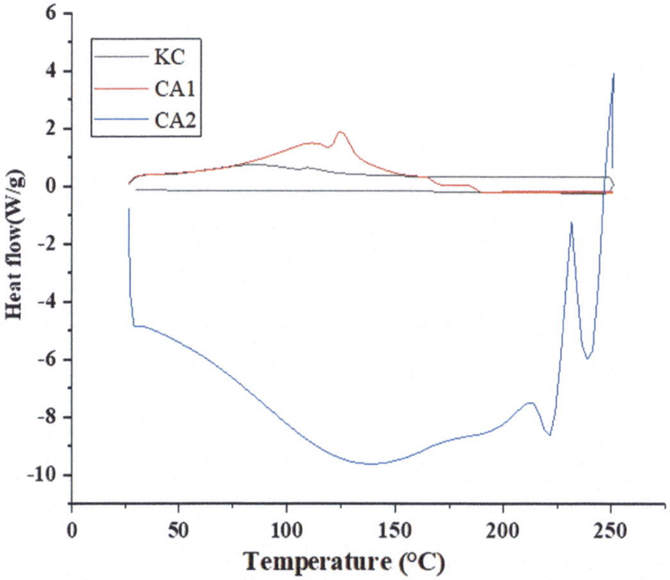

Fig. 4 DSC analysis of Kombucha Bacterial Cellulose (KBC) and Cellulose Acetate (CA)

5 Conclusion

CA was produced from the acetylation of KBC by using two different acetylating agents: acetic acid (CA1) and acetic anhydride (CA2). A total of 68 ± 7.2% of CA1 and 82.66 ± 2.3% of CA2 were obtained from pure kombucha cellulose powder. The CA2 showed the presence of an acetyl group (23.65%) and a higher degree of substitution (1.17) than CA1. The FTIR result showed the presence of acetyl group, hydroxyl group, carboxylic acid, and carbonyl group in both the CA samples. XRD showed a lower degree of crystallinity for CA2 as compared to CA1 and KBC, and the thermal characterization revealed the melting temperature of CA2 at 220 °C, which showed better thermal stability than CA1. The future scope of this study will involve further optimization of the synthesis process by using iodine as a catalyst for a higher degree of substitution and yield of CA. The production of cellulose derivatives, such as cellulose acetate from KBC, is highly desirable due to its biodegradability, environmental sustainability, and economic viability.

Declarations

Ethical Approval

This declaration is not applicable.

Competing Interests

The authors declare that they have no known competing financial interests or personal relationships that could have appeared to influence the work reported in this paper.

Authors' contributions

Laxmi Priya Swain: Conceptualization, Methodology, Investigation, Data curation, Formal analysis, Visualization, Validation, Writing - Original Draft.

Muhil Raj Prabhakar: Conceptualization, Methodology, Investigation.

Balasubramanian Paramasivan: Supervision, Resources, Funding acquisition.

Bhaskar Das: Resources and Mentoring.

Funding

This declaration is not applicable.

Availability of data and materials

All data generated or analyzed during this study are included in this published article.

Acknowledgements The authors are grateful to the Department of Biotechnology and Medical Engineering, NIT Rourkela, for providing their research facilities.

References

1. Ahmad M, Shahir S, Pauliena S, Bohari M, Hasraf N, Nayan M, Izwan S, Razak A (2020) A review on the properties of electrospun cellulose acetate and its application in drug delivery systems: a new perspective. Carbohyd Res 491(February):107978. https://doi.org/10.1016/j.carres.2020.107978
2. Amaral HR, Cipriano DF, Santos MS, Schettino MA, Ferreti JVT, Meirelles CS, Pereira VS, Cunha AG, Emmerich FG, Freitas JCC (2019) Production of high-purity cellulose, cellulose acetate and cellulose-silica composite from babassu coconut shells. Carbohyd Polym 210(January):127–134. https://doi.org/10.1016/j.carbpol.2019.01.061
3. Amarasekara AS, Wang D, Grady TL (2020) A comparison of kombucha SCOBY bacterial cellulose purification methods. SN Appl Sci 2(2):1–7. https://doi.org/10.1007/s42452-020-1982-2
4. Barud HS, de Araújo Júnior AM, Santos DB, de Assunção RMN, Meireles CS, Cerqueira DA, Rodrigues Filho G, Ribeiro CA, Messaddeq Y, Ribeiro SJL (2008) Thermal behavior of cellulose acetate produced from homogeneous acetylation of bacterial cellulose. Thermochim Acta 471(1–2):61–69. https://doi.org/10.1016/j.tca.2008.02.009
5. Beraich FZ, Bakasse M, Arouch M (2017) Synthesis and characterization of cellulose acetate extracted from paper waste. In: Proceedings of 2016 international renewable and sustainable energy conference, IRSEC 2016, pp 785–789. https://doi.org/10.1109/IRSEC.2016.7983970
6. Bertsch P, Etter D, Fischer P (2021) Transient: In situ measurement of kombucha biofilm growth and mechanical properties. Food Funct 12(9):4015–4020. https://doi.org/10.1039/d1fo00630d
7. Blanco Parte FG, Santoso SP, Chou CC, Verma V, Wang HT, Ismadji S, Cheng KC (2020) Current progress on the production, modification, and applications of bacterial cellulose. Crit Rev Biotechnol 40(3):397–414. https://doi.org/10.1080/07388551.2020.1713721

8. Candido RG, Gonçalves AR (2016) Synthesis of cellulose acetate and carboxymethylcellulose from sugarcane straw. Carbohyd Polym 152:679–686. https://doi.org/10.1016/j.carbpol.2016.07.071
9. Cao L, Luo G, Tsang DCW, Chen H, Zhang S, Chen J (2018) A novel process for obtaining high quality cellulose acetate from green landscaping waste. J Clean Prod 176:338–347. https://doi.org/10.1016/j.jclepro.2017.12.077
10. Chen Y, Mastalerz M, Schimmelmann A (2012) Characterization of chemical functional groups in macerals across different coal ranks via micro-FTIR spectroscopy. Int J Coal Geol 104:22–33. https://doi.org/10.1016/j.coal.2012.09.001
11. Cheng HN, Dowd MK, Selling GW, Biswas A (2010) Synthesis of cellulose acetate from cotton byproducts. Carbohyd Polym 80(2):449–452. https://doi.org/10.1016/j.carbpol.2009.11.048
12. Das AM, Ali AA, Hazarika MP (2014) Synthesis and characterization of cellulose acetate from rice husk: Eco-friendly condition. Carbohyd Polym 112:342–349. https://doi.org/10.1016/j.carbpol.2014.06.006
13. De Souza AG, Kano FS, Bonvent JJ, Dos Santos Rosa D (2017) Cellulose nanostructures obtained from waste paper industry: a comparison of acid and mechanical isolation methods. Mater Res 20:209–214. https://doi.org/10.1590/1980-5373-mr-2016-0863
14. Enders AA, North NM, Fensore CM, Velez-Alvarez J, Allen HC (2021) Functional group identification for FTIR spectra using image-based machine learning models. Anal Chem 93(28):9711–9718. https://doi.org/10.1021/acs.analchem.1c00867
15. Fei P, Liao L, Cheng B, Song J (2017) Quantitative analysis of cellulose acetate with a high degree of substitution by FTIR and its application. Anal Methods 9(43):6194–6201. https://doi.org/10.1039/c7ay02165h
16. Fischer S, Thümmler K, Volkert B, Hettrich K, Schmidt I, Fischer K (2008) Properties and applications of cellulose acetate. Macromol Symp 262(1):89–96. https://doi.org/10.1002/masy.200850210
17. Green DS, Almroth BC, Altman R, Bergmann M, Gündoğdu S, Warrier AK, Boots B, Walker TR, Krieger A, Syberg K (2023) Time to kick the butt of the most common litter item in the world: ban cigarette filters. In: Science of the total environment, vol 865. https://doi.org/10.1016/j.scitotenv.2022.161256
18. Homem NC, Amorim MTP (2019) Synthesis of cellulose acetate using as raw material textile wastes. Mater Today Proc 31(xxxx), S315–S317. https://doi.org/10.1016/j.matpr.2020.01.494
19. Khami S, Khamwichit W, Suwannahong K (2019) Synthesis of cellulose acetate nanofiber (CANF) from bacterial cellulose (BC) incubated from cannery seafood wastewater (CSW) using acetobacter xylinum. ARPN J Eng Appl Sci 14(17):3038–3045
20. Kouadri I, Layachi A, Boubendira K, Amor IB, Hemmami H, Zeghoud S, Seghir BB, Rebiai A (2023) A novel eco-friendly source of cellulose acetate extracted from Astragalus gombo seeds: thermal, structural, and morphological characterization. Biomass Convers Biorefinery 2023:1–8. https://doi.org/10.1007/S13399-023-04568-0
21. Laavanya D, Shirkole S, Balasubramanian P (2021) Current challenges, applications and future perspectives of SCOBY cellulose of Kombucha fermentation. J Clean Prod 295:126454. https://doi.org/10.1016/j.jclepro.2021.126454
22. Lingegowda DC, Kumar JK, Prasad AGD, Zarei M, Gopal S (2013) Ftir spectroscopic studies on cleome gynandra–comparative. Romanian J Biophys 22:137–143. (June 2014)
23. Naomi R, Bt R, Idrus H, Fauzi MB (2020) Plant-vs. Bacterial-derived cellulose for wound healing: a review. October. https://doi.org/10.3390/ijerph17186803
24. Nigam S, Das AK, Patidar MK (2021) Valorization of Parthenium hysterophorus weed for cellulose extraction and its application for bioplastic preparation. J Environ Chem Eng 9(4):105424. https://doi.org/10.1016/j.jece.2021.105424
25. Oliver-Ortega H, Geng S, Espinach FX, Oksman K, Vilaseca F (2021) Bacterial cellulose network from kombucha fermentation impregnated with emulsion-polymerized poly (methyl methacrylate) to form nanocomposite. Polymers 13(4):664. https://doi.org/10.3390/polym13040664

26. Rahmayetty, Sulaiman F (2023) Wastewater from the arenga starch industry as a potential medium for bacterial cellulose and cellulose acetate production. Polymers 15(4):1–15. https://doi.org/10.3390/polym15040870
27. Rodrigues Filho G, Monteiro DS, da Meireles CS, de Assunção RMN, Cerqueira DA, Barud HS, Ribeiro SJL, Messadeq Y (2008) Synthesis and characterization of cellulose acetate produced from recycled newspaper. Carbohydr Polym 73(1):74–82. https://doi.org/10.1016/j.carbpol.2007.11.010
28. Seddiqi H, Oliaei E, Honarkar H, Jin J, Geonzon LC, Bacabac RG, Klein-Nulend J (2021) Cellulose and its derivatives: towards biomedical applications. Cellulose 28(4). Springer Netherlands. https://doi.org/10.1007/s10570-020-03674-w
29. Wang J, Tavakoli J, Tang Y (2019) Bacterial cellulose production, properties and applications with different culture methods–a review. Carbohyd Polym 219(May):63–76. https://doi.org/10.1016/j.carbpol.2019.05.008

Automated Identification and Classification of Polymeric Waste Wrappers Using Convolutional Neural Networks and IoT-Based GSM Communication

Kanchan Kumari, Abhishek Barua, and Aman Patra

Abstract Polymeric waste, especially plastics, is a pressing global environmental challenge. This pioneering research amalgamates computer vision, ResNet50 architecture, ReLU activation functions, and IoT communication to transform polymeric waste detection and management. Leveraging Convolutional Neural Networks (CNNs) like ResNet50, the model achieves outstanding accuracy, surpassing 98.7% in identifying diverse polymeric waste items. The integration of ReLU activation functions enhances the network's ability to discern intricate waste patterns, resulting in an impressive F1 score exceeding 98%. Training with an extensive and meticulously annotated dataset ensures adaptability in recognizing varied polymeric waste forms, guaranteeing high accuracy in real-world applications. The IoT integration through a GSM module has played a pivotal role in enabling real-time communication upon waste detection. Recent successful tests showcased the system's seamless transmission of messages from the detection setup to a control room upon polymeric waste identification. Specifically, real-time wrapper detection illustrated the system's efficacy by promptly notifying the designated control room, showcasing its practical applicability and swift responsiveness. This fusion of computer vision and IoT technologies establishes a versatile framework applicable across diverse environments, reshaping waste management in urban areas, public spaces, and recycling facilities. Its scalability positions it as a promising solution to curb polymeric waste proliferation, proactively reducing the ecological footprint of plastics and fostering environmental conservation. This research introduces a transformative paradigm in polymeric waste management. The integration of advanced technologies not only

K. Kumari · A. Patra (✉)
Department of Mechanical Engineering, Parala Maharaja Engineering College, Berhampur, Odisha, India
e-mail: patra.aman12345@gmail.com

A. Barua
Department of Advanced Materials Technology, CSIR-Institute of Minerals and Materials Technology, Bhubaneswar, Odisha, India

Academy of Scientific and Innovative Research, CSIR-HRD Centre Campus, Ghaziabad, Uttar Pradesh, India

© The Author(s), under exclusive license to Springer Nature Singapore Pte Ltd. 2024
M. Suar et al. (eds.), *Advancements in Polymer Technology*, Springer Proceedings in Materials 57, https://doi.org/10.1007/978-981-97-6333-7_6

accurately identifies and categorises waste but also triggers immediate responses, marking a significant stride towards efficient and sustainable waste management practices globally. The successful incorporation of IoT via GSM communication, enabling real-time alerts to a control room, highlights the system's practicality and readiness for immediate integration into waste management infrastructures.

Keywords Automation · Environment IoT · Polymeric waste · Waste management

1 Introduction

Plastic materials, due to their unparalleled versatility, durability, and affordability, have become ubiquitous in modern society. They are the cornerstones of various industries, from packaging and construction to healthcare and electronics. However, this widespread use has led to a pervasive environmental crisis, with plastic waste choking landfills, polluting oceans, and harming wildlife [1–3]. Tackling this escalating challenge requires innovative solutions that address the root problem—efficient identification and management of polymeric waste. In this context, the integration of image processing-based AI models represents a groundbreaking paradigm shift in the realm of waste management. The convergence of cutting-edge technologies, particularly the utilization of the ResNet50 architecture and the Rectified Linear Unit (ReLU) activation function, signifies a monumental leap in computer vision. ResNet50, with its deep neural network structure comprising skip connections, revolutionizes the capacity to discern and categorize polymeric waste with unprecedented accuracy [4, 5]. Its ability to learn intricate features and patterns within images has proven instrumental in distinguishing various types of plastic materials, including wrappers and diverse plastic forms, setting a new standard in waste detection methodologies [6].

The incorporation of ReLU activation functions within these models further enhances their efficacy. ReLU's characteristic of introducing non-linearity by zeroing out negative values significantly amplifies the model's ability to capture nuanced features inherent in plastic waste. This activation function not only accelerates learning but also ensures more accurate identification, contributing substantially to the reliability and precision of waste classification [7]. Furthermore, the amalgamation of IoT-based communication mechanisms, notably GSM modules, complements these AI-driven detection models, enhancing the efficiency of waste management protocols. The integration of IoT enables real-time communication, offering swift and seamless transmission of information upon the detection of polymeric waste. The agility and immediacy of GSM-based communication serve as an efficient conduit for relaying critical data, facilitating prompt alerts, and enabling immediate responses upon identifying plastic waste [8]. The significance of this technological fusion extends beyond mere waste detection—it represents a comprehensive approach to sustainable waste management. The capacity to accurately identify and classify polymeric waste using ResNet50 and ReLU activation functions, combined with efficient IoT-based communication facilitated by GSM modules, revolutionizes the landscape of waste handling practices. It introduces a transformative framework that not only streamlines waste detection processes but also facilitates rapid and informed decision-making, laying the foundation for more sustainable and proactive waste management strategies [9].

As environmental concerns intensify, the application of these advanced technologies offers hope for mitigating the adverse impacts of plastic waste. The ability to swiftly and accurately identify polymeric waste items, coupled with immediate communication and response mechanisms, holds promise for minimizing the environmental footprint of plastic materials. It paves the way for more efficient resource allocation, recycling initiatives, and ultimately, a more sustainable future [10]. In conclusion, the integration of image processing-based AI models, ResNet50 architecture, ReLU activation functions, and IoT-enabled communication systems like GSM modules marks a pivotal moment in waste management evolution. This holistic approach signifies not just a technological advancement but a transformative shift towards more effective, data-driven, and sustainable practices for addressing the global challenge of polymeric waste.

Usha and Mahesh [11] employed the Yolo real-time object detection algorithm to create a smart waste management system based on a deep learning model that optimizes trash isolation and allows for bin status monitoring in an IoT context. They reported that the garbage detection system is more precise and faster than the other existing methods. Flores and Tan [12] discussed various ML techniques for automated segregation of waste and stated their pros and cons, which can help in developing new knowledge systems. Shreya et al. [13] evaluated municipal waste segregation and management processes using IoT-based models and reported efficient waste handling from the perspective of reduced costs, collection time, and optimized routes. Sowmyashree et al. [14] employed the CNN algorithm to segregate the items to be disposed of into bins as plastics and non-plastics. Further, waste is monitored using an ultrasonic sensor, and after the limit is reached, the respective authorities are notified using an application. Karthikeyan et al. [15] developed SSD-based waste separation in smart garbage using augmented clustering NMS and reported that a single-shot object detector effectively detected the waste with the new enhanced technique. Sundaralingam and Ramanathan [16] developed a household waste segregator using the Arduino microcontroller and TensorFlow object detection model and reported that the system can predict the waste class and segregate it into their specific dustbin with mean average precision (mAP) and recall of 86.5% and 88.3%, respectively. Mahajan et al. [17] used UAV movement along with a deep learning algorithm to detect garbage, which helps identify places contaminated with waste and sends an alert message to the waste management system. Bommi et al. [18] proposed an IoT-based smart healthcare waste segregator that separates waste according to the requirement, and waste that needs to be incinerated is burned in the incinerator chamber available in the system. Sigongan et al. [19] developed solar-powered smart garbage segregation bins with SMS notification and machine learning image processing and recommended an algorithm to determine multiple garbage bags at a time. Zubair et al. [20] used AI and mechanical sorting techniques for waste segregation, which can be implemented for commercial waste management purposes. Ullah et al. [21] examined the role of the IoT and ML in managing waste in smart cities and reported their potential to transform urban environments and enable the creation of more sustainable, and efficient cities. Bautista et al. [22] developed an Arduino-based waste detector with an alarm system and reported good efficiency in waste detection, proper waste segregation, and determining if one of the bins is full.

Although lots of work has been done on waste segregation using different techniques such as IoT, AI, ML, ANN, etc.,a few gaps are yet to be filled, such as the segregation of different metals, plastics, polymer compounds, etc. Also, practical implementation of such derived methods requires more time and money. To start segregation in the initial phase, a more simple process can be developed. To address such gaps, this study deals with a transformative paradigm in polymeric waste management. The integration of advanced technologies not only accurately identifies and categorizes waste but also triggers immediate responses, marking a significant stride towards efficient and sustainable waste management practices globally. The successful incorporation of IoT via GSM communication, enabling real-time alerts to a control room, highlights the system's practicality and readiness for immediate integration into waste management infrastructures.

2 Methodology

2.1 Dataset Collection and Preprocessing

Dataset collection and preprocessing are foundational for training accurate models in polymeric waste detection. The dataset, curated from diverse sources including urban, natural, and industrial areas, was enriched through video-to-frame conversion, rapidly generating approximately 60 images per second. A significant portion of the meticulously annotated dataset, comprising over 5000 images, was sourced from reputable repositories like Kaggle, supplementing [23] the research effort with diverse plastic waste representations.

During dataset annotation, rigorous quality control measures were implemented to maintain accuracy and consistency. Clear annotation guidelines were established, providing precise instructions to annotators for labeling various polymeric waste items. This comprehensive dataset encompassed various plastic materials, meticulously annotated to empower the model to discern and categorizes waste accurately during training and inference. Its diversity mirrors real-world scenarios, enabling the model to differentiate between different plastic waste types and conditions across environments. This dataset's role was pivotal in achieving exceptional model performance, reflected in high precision, recall, F1-score, accuracy, and minimal test loss. Its breadth minimized biases, enhanced generalizability, and mitigated overfitting risks, enabling informed predictions in novel waste management contexts.

Fig. 1 Input image

The collection, encompassing diverse sources and Kaggle datasets, along with preprocessing techniques like video-to-frame conversion, formed a robust foundation for training the model in accurate polymeric waste detection. This diverse dataset equips the model to effectively address the environmental challenge of plastic waste. Figure 1 showcases the dataset's variability and complexity through an input image illustration.

2.2 Model Architecture Selection

The selection of a model architecture in the realm of Convolutional Neural Networks (CNNs) and Computer Vision (CV) plays a pivotal role in determining the effectiveness and performance of a system. Among the multitude of architectures available, ResNet-50 stands out as a significant milestone due to its profound impact on image recognition tasks, particularly in the context of plastic detection within computer vision. ResNet-50, an evolution of the Residual Network (ResNet) family, represents a deeper and more sophisticated architecture comprising 50 layers. Its design introduces a fundamental concept known as residual learning, which addresses the challenge of training very deep neural networks. Residual learning incorporates shortcut connections or skip connections within the network's building blocks, enabling the flow of information through alternate paths. This mechanism mitigates the vanishing gradient problem, allowing for the successful training of significantly deeper networks. In the domain of plastic detection through computer vision, ResNet-50 emerges as a robust choice for several reasons. Its architecture facilitates the extraction of intricate features from images, a critical aspect when discerning between various types of plastics, their forms, colours, textures, and sizes. The depth of ResNet-50 enables the model to learn hierarchical representations of these features, capturing both low-level and high-level details present in plastic objects.

One of the distinguishing factors of ResNet-50 lies in its utilization of bottleneck blocks. These blocks employ a combination of 1×1, 3×3, and 1×1 convolutions, reducing computational complexity while retaining the network's depth.

This architecture not only enhances the learning of discriminative features but also optimizes computational efficiency, a crucial consideration for real-time or resource-constrained applications. Furthermore, the incorporation of Global Average Pooling (GAP) instead of traditional fully connected layers at the end of the network aids in reducing overfitting and model complexity. GAP condenses the spatial information within feature maps, allowing the model to generate compact and generalized representations of the input, which is particularly advantageous in scenarios where diverse plastic items need to be accurately classified. ResNet-50's success in plastic detection within CV is underpinned by its ability to learn rich feature representations, its depth, and its capacity to effectively handle complex visual data. Its performance in image classification tasks, including those encountered in plastic detection, has been extensively validated, making it a preferred choice in various real-world applications. ResNet50's convolutional layers follow a structured process to extract intricate features from input images. Beginning with initial convolution and pooling, the network incorporates 16 residual blocks organized into stages. Each block contains convolutional layers, batch normalization, ReLU activation, and crucially, skip connections, facilitating efficient gradient flow for deep learning. The residual connections enable the network to learn residuals between input and output, addressing the vanishing gradient issue. After passing through these blocks, global average pooling reduces spatial dimensions, and a fully connected layer yields the final output. The process flow, depicted in Fig. 2, illustrates the sequence of operations, emphasizing the significance of residual connections in training deep neural networks effectively. This architecture allows ResNet50 to capture complex hierarchical features, contributing to its success in image recognition tasks.

The validation methodology ensured unbiased model evaluation for ResNet50. Cross-validation techniques and diverse validation set compositions were employed. Cross-validation divided the dataset into k folds, iteratively training the model on k-1 folds and evaluating on the remaining fold. This approach provided comprehensive evaluation across different dataset subsets. Additionally, the validation set comprised diverse polymeric waste scenarios from various locations and environmental conditions, enhancing the model's generalizability and effectiveness in real-world settings.

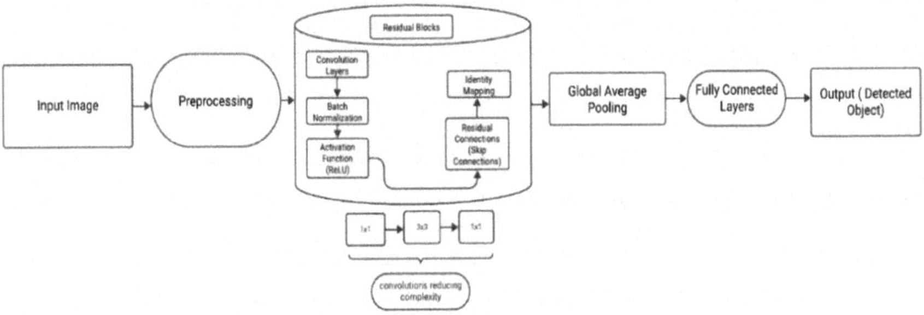

Fig. 2 ResNet50 convolutional layer process flow

Ultimately, the significance of ResNet-50 in the CNN and CV domains for plastic detection stems from its prowess in learning intricate features, handling deeper architectures, optimizing computational resources, and exhibiting exceptional performance in image classification tasks. Its adaptability, accuracy, and efficiency make it a compelling choice for addressing the challenges posed by plastic detection within computer vision applications.

2.3 Activation Function

ReLU was selected for the activation function in this research due to its inherent properties that align with the intricacies of plastic detection within computer vision. When discerning between various types of plastics and their diverse forms, colours, and textures, the choice of activation function significantly influences a model's capability to extract meaningful features. ReLU, short for Rectified Linear Unit, introduces a crucial non-linearity to the neural network architecture. Its simplicity and computational efficiency make it an ideal choice for processing visual data. This activation function zeros out negative values in the input, allowing positive values to pass through unchanged. In the context of plastic detection, this mechanism facilitates the extraction of essential features present in plastic objects while effectively handling variations in lighting, angles, and shapes.

The distinctive characteristic of ReLU—allowing positive values to flow through without constraint—aids in capturing crucial visual cues embedded in images of plastic items. This non-saturating activation function enables the network to learn complex patterns and discriminative features, crucial when distinguishing between different types of plastics or recognizing specific plastic attributes.

Furthermore, ReLU helps mitigate the vanishing gradient problem often encountered in deeper neural networks. This property becomes particularly advantageous in the context of plastic detection, where deeper architectures are required to comprehend the diverse and intricate characteristics of plastic materials. By preventing the saturation of gradients, ReLU enables the network to learn more effectively from the complex visual data related to plastics. In the realm of computer vision, the choice of activation function significantly influences a model's capacity to discern relevant features from images. ReLU's ability to introduce non-linearity, handle complex visual data, and facilitate effective feature extraction aligns with the demands of plastic detection. Its inherent advantages in promoting the learning of discriminative features, handling deeper architectures, and addressing gradient vanishing make it a suitable and advantageous choice for enhancing the accuracy and efficiency of plastic detection models within computer vision frameworks.

The Rectified Linear Unit (ReLU) activation function, denoted as $f(x) = \max(0,x)$, operates by setting negative input values to zero and retaining positive values unchanged, as shown in Fig. 3. Mathematically, when applied to the input x, the function output f(x) can be represented as follows:

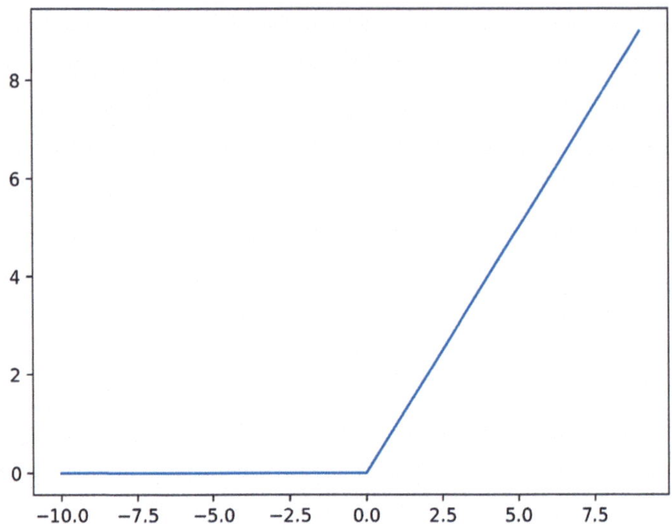

Fig. 3 Graph representing ReLU [24]

$$f(x) = \begin{cases} x \; ifx > 0 \\ 0 \; ifx \leq 0 \end{cases} \tag{1}$$

2.4 *Evaluation Metrics*

In Convolutional Neural Networks (CNNs) applied to polymeric waste classification, evaluation metrics are crucial to assessing the model's performance. Metrics like accuracy, precision, recall, F1 score, and support are instrumental in gauging the effectiveness of the model's predictions. Accuracy measures the proportion of correctly classified polymeric waste items over the total number of items classified. It provides an overall view of the model's correctness but may not be sufficient when classes are imbalanced. Precision indicates the accuracy of positive predictions, showcasing the proportion of correctly identified polymeric waste items among all items classified as waste. It's vital in scenarios where false positives should be minimized. Recall, also known as sensitivity, illustrates the model's ability to identify all relevant instances of polymeric waste, calculating the proportion of correctly identified waste items among all actual waste items. The F1 score is a harmonic mean of precision and recall, providing a balanced measure between the two. It offers a single metric that considers both false positives and false negatives, making it valuable in assessing the model's overall performance. Support indicates the number of actual occurrences of each class in the dataset. It aids in understanding the dataset's distribution and how well the model performs on different classes.

In the assessment of polymeric waste classification, these metrics were applied rigorously. Accuracy helped gauge the overall correctness of waste identification. Precision and recall were crucial, as minimizing false positives and capturing as many waste items as possible were paramount. The F1 score provided a balanced evaluation, considering both precision and recall. Finally, support aided in understanding the distribution of different waste classes and how well the model performed on each type of waste item. By employing these evaluation metrics, the CNN model's performance in accurately classifying polymeric waste items was comprehensively assessed, enabling insights into its effectiveness and guiding potential improvements in waste detection and management systems.

2.5 *IoT Integration and* Communication *Protocols*

In the realm of waste management, the integration of IoT and communication protocols, exemplified by the Arduino and GSM modules, revolutionizes the efficiency of polymeric waste detection systems. The amalgamation of these technologies enables seamless communication between an AI-powered waste detection system and external networks, paving the way for rapid alerts and notifications upon the identification of polymeric waste. Central to this integration is the Arduino, which functions as a pivotal component for orchestrating communication between disparate elements of the system. The Arduino, serving as a microcontroller, facilitates serial communication protocols to establish connectivity between the AI-based detection system and a host computer, typically a PC. This interaction enables the transmission of data from the image processing conducted on the PC, leveraging an external webcam to capture images of waste, to the AI algorithms, particularly ResNet-based models, for waste classification and identification. The AI algorithms, residing on the PC, leverage the ResNet architecture to process and classify polymeric waste items captured by the webcam. As the image processing and classification unfold on the PC, the Arduino, through serial communication, acts as an intermediary, facilitating the exchange of information between the AI model and the processing capabilities of the PC.

2.6 *Experimental Setup*

The experimental setup and validation process encompassed a meticulous fusion of hardware and software components to fortify the reliability and robustness of the polymeric waste detection system. At the heart of this configuration were the Arduino microcontroller, a PC embedded with proficient image processing capabilities, ResNet-based AI algorithms, and a GSM module interfacing with an external telephone network.

Figure 4 illustrates the intricate connections between the Arduino and the GSM module, showcasing their pivotal roles in orchestrating seamless communication within the system. The hardware arrangement featured the Arduino as a pivotal communication hub, orchestrating seamless interaction between the PC and the GSM module. Employing serial communication protocols, the Arduino established a robust link with the PC, facilitating the transmission of waste image data captured by an external webcam to the AI algorithms for classification. Simultaneously, the integration of the GSM module into the system was pivotal. Interfacing seamlessly with the Arduino, it facilitated communication with an external telephone network. This integration enabled the real-time transmission of alerts or notifications upon the successful identification of polymeric waste by the AI algorithms. The software infrastructure primarily revolved around the PC, serving as the powerhouse for image processing and execution of the ResNet-based AI algorithms for waste classification. Leveraging the webcam, the PC captured polymeric waste images, initiating the image processing pipeline. The integrated ResNet algorithms meticulously classified the waste items based on processed images, harnessing the neural network's depth and learning capabilities.

The validation of this experimental setup involved comprehensive testing to gauge the system's performance. The diverse polymeric waste image dataset was instrumental in evaluating the AI algorithms' accuracy, precision, recall, F1 score, and support metrics. This extensive testing scrutinised the system's aptitude to accurately identify various types of polymeric waste, ensuring its efficacy in practical scenarios. To fortify the model's reliability, cross-validation techniques were employed. The system underwent testing with distinct dataset subsets, validating its performance across various waste scenarios and environmental conditions. This validation aimed to affirm the model's predictive capabilities, ensuring adaptability while mitigating biases and overfitting tendencies. In conclusion, the meticulous integration of hardware and software components, featuring the Arduino, PC, AI algorithms, and GSM module, ensured a comprehensive and reliable polymeric waste detection system.

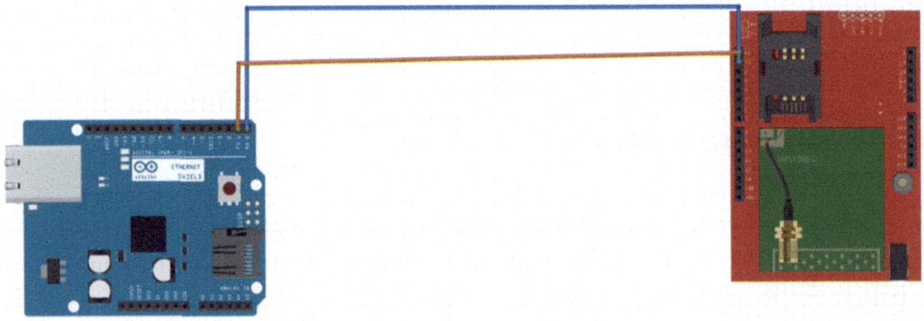

Fig. 4 Arduino and GSM connection

Rigorous testing and validation underscored the system's accuracy, reliability, and adaptability, affirming its potential in real-world waste management applications.

3 Results and Discussions

The polymeric waste detection system showcased exceptional performance across multiple metrics. Precision, standing at 0.976119, highlighted the system's accuracy in identifying plastic bag images among all predicted positive instances. Accompanied by an impressive accuracy score of 0.987915, the system illustrated its overall precision in classifying polymeric waste items, affirming its reliability. Further emphasizing its effectiveness, the F1 score reached an impressive 0.981982, signifying balanced accuracy in discerning and categorizing plastic bag images. Additionally, the system displayed minimal test loss, measuring at 0.03352. This minimal loss underlined the system's capability to minimize discrepancies between predicted and actual values, ensuring highly precise classification of polymeric waste items. Figure 5 illustrates the minimal loss observed with respect to epochs, while Fig. 6 depicts the system's accuracy graph with respect to epochs, showcasing a remarkable 98.7% accuracy. Furthermore, Fig. 7 displays real-time detection of wrappers, illustrating the system's practical applicability. Lastly, Fig. 8 represents the successful receipt of SMS alerts by GSM, validating the system's real-time communication capabilities. These robust results, combined with the visual representations in Figs. 5, 6, 7, and 8, underscore the system's exceptional accuracy, precision, and efficacy in polymeric waste classification. The system's consistent and reliable performance across various metrics and real-time practical applications emphasizes its potential for effective implementation in waste management protocols.

The Arduino interfaces with the GSM module, a crucial element that enables communication with an external telephone network. This GSM module, leveraging cellular networks, establishes a channel for communication from the waste detection system to designated mobile devices or systems configured to receive alerts or notifications. Upon the successful identification of polymeric waste by the ResNet-based AI algorithms, triggered by the webcam images processed on the PC, the Arduino promptly triggers the GSM module. This activation prompts the transmission of immediate alerts or notifications to predetermined recipients via the external telephone network. These alerts serve as real-time notifications, providing crucial information about the detected presence of polymeric waste, thereby facilitating swift responses and interventions in waste management protocols. This seamless integration of the Arduino, PC-based image processing, ResNet-based AI algorithms, and the GSM module signifies a paradigm shift in waste detection and communication methodologies. The interconnectedness of these components establishes a responsive and efficient system capable of identifying and communicating the presence of polymeric waste in real time, enabling proactive waste management strategies. By leveraging IoT technologies and communication protocols, this integrated system not only enhances waste detection accuracy but also enables timely interventions, laying

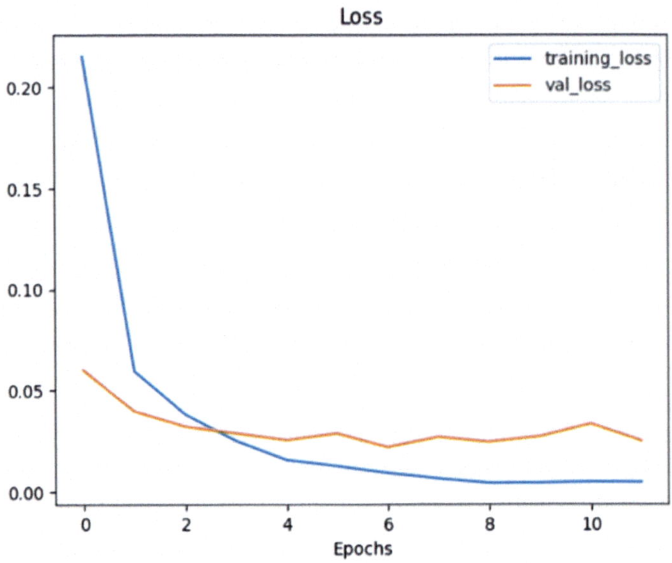

Fig. 5 Graph representing losses

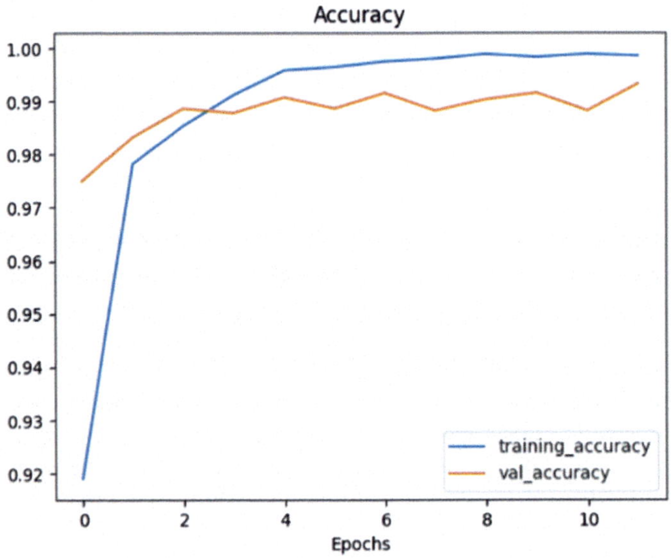

Fig. 6 Graph representing accuracy

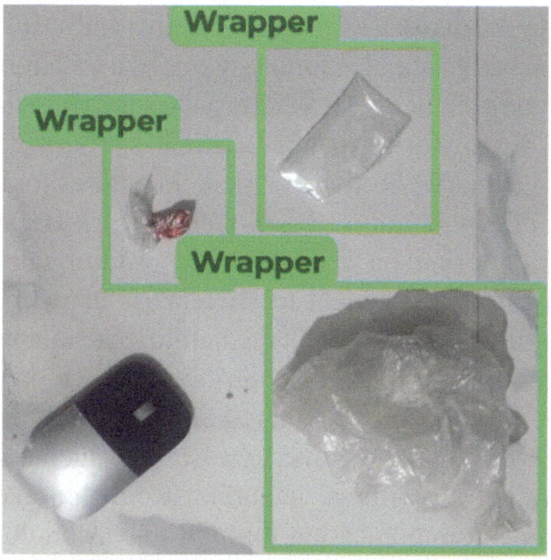

Fig. 7 Output, wrappers getting detected

Fig. 8 SMS alert via GSM

the groundwork for more effective and sustainable waste management practices. This convergence of technology marks a transformative step towards addressing the escalating challenges posed by polymeric waste in environmental conservation endeavors. Looking forward, the research has promising future prospects in advancing computer vision and neural network architectures for even higher accuracy in waste identification. Integration of IoT technologies is expected to expand, facilitating real-time monitoring on a larger scale. Additionally, the application could extend beyond waste management to encompass broader environmental monitoring and sustainability efforts. With its scalability and adaptability, these technologies hold potential for diverse contexts, contributing to a more sustainable future.

The novelty of this study lies in its integration of advanced technologies to revolutionize polymeric waste management. By combining cutting-edge computer vision algorithms, including Convolutional Neural Networks (CNNs), with real-time communication through IoT technologies, the system offers a transformative approach to waste detection and classification. This integration enables swift and accurate identification of polymeric waste items, especially plastic bags, addressing a critical environmental challenge. The seamless fusion of these technologies presents a novel solution that enhances efficiency and accuracy in waste management practices, promising significant advancements in sustainable waste handling.

Implementing the proposed waste management system in India entails regulatory and policy considerations, notably concerning data privacy and compliance. Adhering to laws like the Personal Data Protection Bill, 2019, is crucial for safeguarding individuals' information. Compliance with environmental regulations such as the Environmental Protection Act, 1986, and Plastic Waste Management Rules, 2016, is vital for sustainable waste management. Collaboration with bodies like the Central Pollution Control Board and stakeholder engagement is necessary for alignment with existing policies and public acceptance. Addressing these concerns ensures smooth integration into India's waste management infrastructure.

4 Conclusion

This research introduces a transformative paradigm in polymeric waste management. The integration of advanced technologies not only accurately identifies and categorizes waste but also triggers immediate responses, marking a significant stride towards efficient and sustainable waste management practices globally. The successful incorporation of IoT via GSM communication, enabling real-time alerts to a control room, highlights the system's practicality and readiness for immediate integration into waste management infrastructures.

In conclusion, the polymeric waste detection system has proven itself a highly accurate and efficient solution. With precision at 0.976119 and an accuracy score of 0.987915, it excels at identifying plastic bag images among various waste items. The F1 score of 0.981982 demonstrates its balanced accuracy in categorizing plastic waste. Furthermore, the system's minimal test loss of 0.03352 reflects its ability to minimize errors, ensuring precise waste classification. Visual representations in Figs. 5, 6, 7, and 8 validate its robustness, highlighting minimal loss, high accuracy, real-time detection, and successful communication via GSM. These results emphasize the system's effectiveness in waste classification, particularly in identifying plastic bags, highlighting its potential for real-world waste management. Its consistent performance across metrics presents a promising solution for addressing environmental challenges posed by plastic waste, contributing to more sustainable waste handling practices.

Overall, the future scope of automated identification and classification of polymeric waste wrappers using CNNs and IoT-based GSM communication holds promise for improving waste management practices, reducing environmental pollution, and fostering a more sustainable future. Some potential future developments and implications are enhanced waste management efficiency, real-time monitoring and data analytics, scalability and adaptability, remote management and control, integration with smart waste bin, environmental impact assessment, education and awareness, potential for innovation and collaboration, etc.

References

1. Sio GA, Guantero D, Villaverde J (2022) Plastic waste detection on rivers using YOLOv5 algorithm. In: 2022 13th International conference on computing communication and networking technologies (ICCCNT). IEEE, pp 1–6
2. Liu L, Zhou B, Liu G, Lian D, Zhang R (2022) Yolo-based multi-model ensemble for plastic waste detection along railway lines. In: IGARSS 2022–2022 IEEE international geoscience and remote sensing symposium. IEEE, pp 7658–7661
3. Wu W, Liu H, Li L, Long Y, Wang X, Wang Z, Li J, Chang Y (2021) Application of local fully convolutional neural network combined with YOLO v5 algorithm in small target detection of remote sensing image. PLoS ONE 16(10):e0259283
4. Ishikawa C, Marasigan JAU, Caya MVC (2020) Cloud-based signature validation using CNN inception-ResNet architecture. In: 2020 IEEE 12th international conference on humanoid, nanotechnology, information technology, communication and control, environment, and management (HNICEM). IEEE, pp 1–6
5. Schmidt C, Krauth T, Wagner S (2017) Export of plastic debris by rivers into the sea. Environ Sci Technol 51(21):12246–12253
6. Si J, Harris SL, Yfantis E (2018) A dynamic ReLU on neural network. In: 2018 IEEE 13th Dallas circuits and systems conference (DCAS). IEEE, pp 1–6. (Nov 2018)
7. Yang P, Dong C, Zhao X, Chen X (2020) The surface damage identifications of wind turbine blades based on ResNet50 algorithm. In: 2020 39th Chinese control conference (CCC). IEEE, pp 6340–6344
8. Karmokar C, Hasan J, Khan SA, Alam MII (2018) Arduino UNO based smart irrigation system using GSM module, soil moisture sensor, sun tracking system and inverter. In: 2018 International conference on innovations in science, engineering and technology (ICISET). IEEE, pp 98–101
9. Barkat HSE, Mahdad B (2018) An experimental educational platform based Arduino-GSM for power system protection. In: 2018 International conference on communications and electrical engineering (ICCEE). IEEE, pp 1–6

10. Cho H, Choi HJ, Lee CE, Sir CW (2019) Sitting posture prediction and correction system using arduino-based chair and deep learning model. In: 2019 IEEE 12th conference on service-oriented computing and applications (SOCA). IEEE, pp 98–102
11. Usha SM, Mahesh HB (2022) Accurate and high speed garbage detection and collection technique using neural network and machine learning. In: IOP conference series: materials science and engineering, vol 1258, no 1, p 012055. IOP Publishing
12. Flores MG, Tan J (2019) Literature review of automated waste segregation system using machine learning: a comprehensive analysis. Int J Simul Syst Sci Technol 11
13. Shreya M, Katyal J, Ramesh R (2023) Technical solutions for waste classification and management: a mini-review. Waste Manag Res 41(4):801
14. Sowmyashree MS, Saritha IG, Abhilash S, Sharma A, Saravanan S, Rao SN (2022) Machine learning based reward bin for effective waste segregation. In: AIP conference proceedings, vol 2463, no 1. AIP Publishing
15. Karthikeyan M, Subashini TS, Jebakumar R (2021) SSD based waste separation in smart garbage using augmented clustering NMS. Autom Softw Eng 28:1–17
16. Sundaralingam S, Ramanathan N (2023) A deep learning-based approach to segregate solid waste generated in residential areas. Eng Technol Appl Sci Res 13(2):10439–10446
17. Mahajan S, Mulla AFN, Patil R, Salunkhe M, Patil P, Gejage A (2023) Drone based survey of garbage locations for waste management. In: AIP conference proceedings, vol 2717, no 1. AIP Publishing
18. Bommi RM, Rajeev SVS, Navya S, Teja VS, Supriya U (2023) Smart health care waste segregation and safe disposal. Math Comput Sci 2:205–221
19. Sigongan JB, Sinodlay HP, Cuizon SXP, Redondo JS, Macapulay MG, Bulahan-Undag CO, Gumonan KMVC (2023) GULP: solar-powered smart garbage segregation bins with SMS notification and machine learning image processing. arXiv:2304.13040
20. Zubair M, Mathur Y, Rathore H, Gupta P, Banerjee S (2022) Smart waste bin: mechanical and AI based waste segregation. In: 2022 Second international conference on artificial intelligence and smart energy (ICAIS). IEEE, pp 63–66
21. Ullah A, Anwar SM, Li J, Nadeem L, Mahmood T, Rehman A, Saba T (2023) Smart cities: the role of Internet of Things and machine learning in realizing a data-centric smart environment. Complex Intell Syst 1–31
22. Bautista MXL, Aranas CB, Arcigal RJ (2023) Arduino-based waste detector with alarm system. Eur J Sci Innov Technol 3(1):146–158
23. Vencerlanz09 (2022) Plastic paper garbage bag synthetic images dataset. Kaggle. https://www.kaggle.com/datasets/vencerlanz09/plastic-paper-garbage-bag-synthetic-images/data
24. Asker B, Al-Allaf A (2022) Detecting cheating in electronic exams using the artificial intelligence approach. 7:999–1007

Reduction of Formaldehyde Emission from Plywood Panels Bonded with Phenol-Formaldehyde Resin (PF) by Using Polymate-777P as a Formaldehyde Scavenger

S. C. Sahoo, Shibu Comath, and Chinmoya Das

Abstract Formaldehyde emissions from wood composites, commonly used in wood-based panel industries, pose significant environmental and health concerns. This research aimed to assess the effectiveness of Polymate-777P, a commercially branded formaldehyde scavenger, in mitigating formaldehyde emissions from plyboard manufactured using phenolic resin as a binder. The study involved a comprehensive evaluation of formaldehyde emission reduction in wood composites. A series of experiments were conducted to investigate the impact of Polymate-777P by using 0.5, 1.0, 1.5, 2.0, and 3.0% concentrations of the ply board bonded with phenol formaldehyde resin. The impact of the scavenger Polymate-777P on the properties of Phenol formaldehyde resin, including parameters such as viscosity, pH, gelation time, and free formaldehyde content, was assessed. Additionally, an evaluation of the physico-mechanical properties and formaldehyde emission of the resulting plywood was conducted. The findings of the study demonstrate that the manufactured plywood met the Japanese industrial standard F*** emission requirement ($\leq$0.50 mg/L) and European E_0 classes when Polymate-777P was employed as a scavenger at concentrations of 1.5% or higher. The study also revealed that scavenger not only reduces the formaldehyde emission of the ply board but also increases the bonding properties and technological properties of the board to some extent. The study shows the feasibility of using Polymate-777P as a formaldehyde scavenger and offers a promising solution to minimize formaldehyde emissions, contributing to improved environmental and health outcomes in wood composite industries.

S. C. Sahoo (✉)
Institute of Wood Science and Technology (Indian Council of Forestry Research & Education), Field Station, Kolkata, West Bengal, India
e-mail: sahoosc@icfre.org

S. Comath
College of Forestry, Department of Forest Products and Utilization, Kerala Agricultural University, Thrissur, Kerala, India

C. Das
Patina Chemicals, Amtala, 24 Progonas South, Amtala, West Bengal, India

© The Author(s), under exclusive license to Springer Nature Singapore Pte Ltd. 2024
M. Suar et al. (eds.), *Advancements in Polymer Technology*, Springer Proceedings in Materials 57, https://doi.org/10.1007/978-981-97-6333-7_7

Keywords Wood composite · Formaldehyde emission · Formaldehyde scavenger · Phenol formaldehyde resin · Polymate-777P

1 Introduction

Formaldehyde is a volatile organic compound (VOC) commonly emitted by wood-based materials, including plywood, particleboard, and fiberboard, a known carcinogen to humans and harmful to the environment. The cost-effectiveness and established performance have solidified formaldehyde-based resins as the primary wood adhesive for both interior and exterior applications. However, formaldehyde emissions from wood products have gained recognition as a potential source of indoor air pollution, contributing to discomfort among occupants and posing potential health concerns. Numerous studies have demonstrated the significant potential of various scavengers to mitigate formaldehyde emissions from wood composites [1, 2, 5, 11, 12]. For example, a study by Saito et al. [11] found that 0.20 weight percentage of graphene oxide effectively reduced formaldehyde levels in indoor environments by up to 81%. Similarly, Ashaari et al. [2] showed that 3% of ammonium hydroxide effectively adsorbs formaldehyde molecules and significantly reduces approximately 88% of formaldehyde emissions. Additionally, research is under progress to develop alternative wood adhesives with lower formaldehyde emissions while maintaining the same level of performance and durability. However, it is important to note that excessive formaldehyde scavengers introduced into formaldehyde-based adhesives can compromise internal bond strength and harm product properties.

Formaldehyde scavengers are categorized as synthetic, bio-based, and nano scavengers [6]. The choice of formaldehyde scavenger depends on the adhesive's specific application and desired properties. Synthetic scavengers are commonly used due to their cost-effectiveness and wide availability [2]. Bio-based scavengers, on the other hand, offer a more sustainable and environmentally friendly alternative [4]. Nano scavengers, with their unique properties at the nanoscale, show promising potential for enhancing formaldehyde scavenging efficiency [3].

The gel time (GT) of the Urea Formaldehyde (UF)/Polymeric diisocyanate (pMDI) hybrid resin was shortened when 10% pMDI was added to the liquid UF resin [8]. The study's findings suggest that modifying liquid UF resins with pMDI improves water resistance, and there are no significant changes in the E_2 emission standard compared to the control group [7, 8]. Formaldehyde-free and environment-friendly GL–Corn starch/pMDI-based adhesive could be used for making plywood and panel products at a specific pressure equal to 12–14 kg/cm^2 and temperature of 145–150 °C. Formaldehyde emission content in bonding products as per national standards was used [10]. Even at low concentrations, incorporating nanomaterials as extenders in wood panels can significantly change mechanical properties [9].

In this research, Polymate-777P was incorporated into phenolic resin with the objective of assessing its efficacy as a formaldehyde scavenger. To the best of current knowledge, no prior investigation exists regarding the impact of Polymate-777P incorporation into phenol formaldehyde resin (PF resin) on both formaldehyde emissions and the properties of plywood. Consequently, this paper is dedicated to elucidating the effects of Polymate-777P addition on plywood properties and formaldehyde emissions.

2 Materials and Methods

2.1 Materials

Technical-grade formaldehyde (37%), Phenol (99%), Sodium hydroxide/caustic soda (96%), and aqueous sodium hydroxide (NaOH) procured from the local market were used for the synthesis of PF resins.

Eucalyptus was used as the core, and Gurjan was employed as the surface veneer to produce plywood. Veneers were dried up to 4–6% moisture content before manufacturing the plyboard.

Polymate-777P is a commercial brand additive used as a formaldehyde scavenger supplied by M/S-Patina Chemicals, Kolkata, India. This scavenger is formulated by mixing some specialty chemicals to achieve the E_1 and E_0 emission levels of the wood-based composite board manufactured by Phenol formaldehyde resin.

2.2 Methods

2.2.1 Physical and Chemical Properties Study of the Scavenger Polymate-777P

The physical and chemical properties like color, odor, solubility, pH, ash content, flammability, solid content, moisture content, and chemical compatibility with the polymate-777P were tested.

2.2.2 Preparation of PF Resin

A three-necked 1000 ml round bottom flask was filled with 100 parts of molten phenol (98%), followed by 180 parts by weight of formalin (37%). The flask was equipped with a reflux condenser, a mechanical stirrer, and a digital thermometer, and then 08 parts of caustic soda (50%) aqueous solution was added to the mixture. After stirring for 10–20 min, the gentle heating process began. Given the exothermic nature of the reaction, heating was stopped after exothermicity was established, and the temperature was kept between 90 and 92 °C. The reaction was maintained until

Table 1 The concentration of Polymate-777P adopted for the experiment

Treatments	Scavenger concentration (%)
PF0	0.0
PF1	0.5
PF2	1.0
PF3	1.5
PF4	2.0
PF5	3.0

the water tolerance reached 1:8/9, and the B4 Ford cup had a flow time of about 15–16 s as per IS 3944:1969.

Subsequently, the resin was cooled to 60–70 °C, and Polymate-777P: Water (1:1) was slowly added with continuous stirring, then further cooled to room temperature. Six different resins were synthesized using Polymate-777P as a scavenger, with concentrations varying from 1 to 5%, including a control. The unloaded resin was allowed to undergo complete polymerization and conditioning over 12 h. The synthesized resin was then utilized in plywood manufacturing, and assessments were made for shelf life and free formaldehyde content. The PF resin was prepared in the above procedure by using Scavenger Polymate-777P at various concentrations: 0.0% (PF0-control), 0.5% (PF1), 1.0% (PF2), 1.5% (PF3), 2.0% (PF4) and 3.0% (PF5) respectively.

2.2.3 Experimental Design and Data Analysis

To evaluate the efficacy and possible applications of the scavenger Polymate-777P in conjunction with PF adhesive, a 1 × 6 × 3 factorial experiment was conducted using a completely randomized design (CRD). The experimental design included 1 type of adhesive, 6 levels of scavenger extension, and 3 replicates, resulting in 18 combinations under this study. The extension levels of the scavenger will be 0.5, 1.0, 1.5, 2.0, and 3.0% including blank (0%). Table 1 shows the adopted concentration of Polymate-777P for the experiment.

2.2.4 Properties of PF Resin

The PF resins were characterized using several parameters, including viscosity, solids content, gel time, and free formaldehyde.

Viscosity

A variation in glue viscosity has significant implications for glue spreader rates. Glue mixes with high viscosity pose challenges in spreading, resulting in increased glue spread, which is undesirable both in terms of quality and economic considerations. In the context of resins, viscosity is a metric for evaluating the advancement and degree of polymerization during either the manufacturing process or application.

Within the plywood industry, the actual viscosity of the adhesive is examined using a standard flow cup. In this study, the flow time of the adhesive mix was investigated using the B6 and resin in the B4 Cup as per IS:3944-1982.

Solid Content

A resin sample weighing 5.0 g was placed in an evaporating dish (W_1), and the dish was re-weighed with the resin sample (W_2). The resin was evenly distributed in the dish, placed in a hot air oven set at 105 °C, and left to dry for a duration of 3 h. Subsequently, the dried sample was moved to a desiccator to cool down to room temperature. The dish containing the dried resin was then weighed (W_3). The percentage weight by weight of the resin's solid content was determined using the equation provided below (Eq. 1).

$$\textbf{Solid content}\ (\%) = \frac{\mathrm{W3} - \mathrm{W1(g)}}{\mathrm{W2} - \mathrm{W1}} \times 100 \tag{1}$$

Measurement of Gelation Time

About 5 gm of the resin mix was placed into a test tube, and a thin wire spring was placed in the tube to mix the contents manually. The test tube was positioned in an oil bath maintained at a consistent temperature range of 100–120 °C, and thorough mixing ensued until the liquid resin underwent a transformation into a gel state. The gelation time was recorded.

pH-Value

The pH is monitored automatically in a potentiometric pH meter (Mettler Toledo pH Meter, FP20BIO KIT).

Free Formaldehyde Content in PF Resin

The assessment of free formaldehyde was carried out using the hydroxylamine hydrochloride method (ISO: 9397) with endpoint titration. The modified PF resins

Table 2 Typical glue formulation and panel manufacturing parameters

Material	Concentration
Resin	200 parts
Wheat flour	16 parts
Hardener	0.8 parts
Liq. Ammonia	1.6 ml
Veneer moisture	6–8%
Spread rate	2.8–3.0 m^2/kg
Open assembly time	30 min
Pressing temperature	110 ± 5^0C
Specific pressure	12–14 kg/cm^2
Pressing time	12 min for 9 mm plywood

were dissolved in an isopropanol-water mixture (2:1, v/v). Subsequently, 0.1N hydrochloric acid (HCl) was introduced to the solution under stirring on a magnetic stirrer to adjust the pH to 3.5. In the final step, 10 ml of a hydroxylamine hydrochloride solution (10%, w/w) was incorporated into the solution. After allowing it to react for 10 min, the resulting solution underwent titration with 0.1N sodium hydroxide until reaching a pH of 3.5. The free formaldehyde content in the resin was determined by noting the titration values for both the sample and the blank. This analysis was conducted on all six samples with varying concentrations of 0.0, 0.5, 1.0, 1.5, 2.0, and 3.0% labeled as PF0, PF1, PF2, PF3, PF4, and PF5, respectively.

2.2.5 Manufacturing of Plywood

Plywood with a size of 600 mm × 600 mm × 9 mm was manufactured by using eucalyptus species as core and gurjun (*Dipterocarpus turbinatus*) as face veneer and by using PF resin (Table 2). The adhesive was applied at a rate of approximately 280–300 g/m^2 during the production of 9 mm plywood. The plywood underwent hot-pressing at a temperature range of 140–145 °C using PF resin, applying a specific pressure of 11–12 kg/cm^2. Two plyboard samples were prepared for each experiment to evaluate the technological property. A total of 12 numbers of plywood were prepared to design the experiment.

2.2.6 Determination of Formaldehyde Emission

Perforator Method

The formaldehyde emission perforator value was ascertained in accordance with the DIN EN 120 (Deutsches Institut fur Normung Europaische Norm) method (European Committee for Standardization, 1991). An accurately weighed test piece, amounting to approximately 110 g with a precision of 0.1 g, was placed within a round-bottom flask. Subsequently, 600 ml of toluene was introduced into the flask. The perforator attachment received one liter of distilled water, which was heated to boiling. Boiled toluene was then passed through the distilled water for a duration of two hours. During this process, the distilled water effectively absorbed formaldehyde and other volatile organic compounds (VOCs) liberated by the boiling toluene. The formaldehyde captured by the water was subsequently quantified using an Ultra Violet spectrophotometer after treatment with acetylacetone and ammonium acetate solution at 412 nm.

Desiccator Method

The assessment of free formaldehyde emission followed the desiccator method outlined in the JIS A 1460:2015 standard (Japanese Industrial Standard). This method involves determining the amount of formaldehyde released from plywood specimens, which is subsequently absorbed in a specified volume of distilled water over 24 h within a glass desiccator.

Ten test pieces from each plywood type, each measuring 150 ± 1 mm in length and 50 ± 1 mm in width, were prepared, along with a blank sample. These specimens were then positioned in a desiccator with a sealed volume of 11 L, containing a glass-crystallizing dish filled with 300 mL of distilled water. After 24 h, the samples were extracted from the desiccators, and the resulting formaldehyde solution was readied for spectroscopic analysis.

2.2.7 Evaluation of Plywood Properties

The properties of the plywood encompass the modulus of rupture (MOR), modulus of elasticity (MOE), and glue shear strength (GSS) under dry conditions. Testing was conducted to assess water resistance properties in accordance with IS 1734:1983 standards. Subsequently, each plywood variant underwent comprehensive testing for water resistance properties in accordance with IS: 848-2006, which outlines specifications for synthetic resin adhesives applicable to both Moisture Resistance (MR) and Boiling Water-Proof (BWP) grades.

Table 3 Typical properties of Polymate-777P

Appearance	White in color, free-flowing powder
Odor	Characteristic (Slight)
Solubility	Soluble in water
pH (10% Solution)	6–7
Ash content (%)	<3.0
Flammability	Non-flammable
Solid (%)	>95
Moisture content (%)	<5

3 Result and Discussion

3.1 Typical Properties of Polymate-777P

The physical and chemical properties of the additive Polymate-777P were carried out as per IS 1508:1972 (Table 3). The result shows satisfactory value, which is within the acceptance criteria. Polymate-777P was found compatible with the phenolic resin during manufacturing without creating any adverse effect on the qualities of the resin.

3.2 Effect of Scavenger on Resin Properties

Analyzing the results presented in Table 4, it becomes apparent that there are no significant alterations in the viscosity and solid content of the resin across the various scavenger concentrations. However, an intriguing observation emerges in relation to the gel time of the resin. This parameter demonstrates a consistent decrease as the concentration of scavenger load increases. This decline in gel time has a notable impact on the curing kinetics of the PF resin.

Table 4 Typical properties of prepared PF resin

Sample name	Flow time (S)	Solid content (%)	Gel time (min)	pH	Free formaldehyde (%)
PF0	21	50.2	13.76	10.86	0.62
PF1	20	49.8	13.62	10.89	0.58
PF2	22	49.6	13.54	10.91	0.40
PF3	22	49.7	13.48	10.88	0.34
PF4	24	49.8	13.44	10.82	0.28
PF5	21	49.9	13.40	10.84	0.20

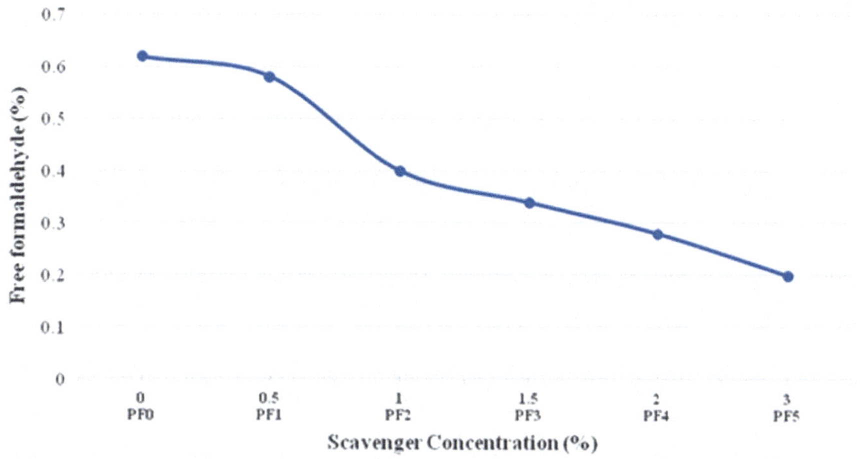

Fig. 1 Free formaldehyde percentage at different scavenger concentrations

Furthermore, the evaluation of free formaldehyde levels within the PF resin, as depicted in Fig. 1, reveals a significant reduction in all cases compared to the reference sample (PF0). This reduction in formaldehyde content within the resin is of considerable importance. It not only signifies an environmentally favorable outcome by reducing the presence of formaldehyde in the resin but also holds the potential to translate into reduced formaldehyde emissions from the wood composite board. This reduction in emissions aligns with the broader goal of producing wood composite boards that are more environmentally friendly and safer for various applications, ultimately enhancing the overall sustainability and quality of the product.

3.3 *Mechanical Properties and Bond Quality*

The static bending properties of the plywood demonstrate a positive correlation with an increasing concentration of the scavenger. It is noteworthy that the specimens with scavenger content ranging from 0.5 to 3% exhibited the highest modulus of rupture values within the plywood. As the scavenger content increased from 0.5% (PF1) to 3% (PF5), the MOR consistently rose, escalating from an initial 54.62 N/mm^2 to a peak of 58.61 N/mm^2. However, it is worth noting that beyond 3%, there was no significant further increase in the MOR values, indicating a potential threshold effect.

The Modulus of Elasticity (MOE) of the plywood also demonstrated an upward trend with increasing scavenger content, as evident in Table 5. Importantly, at 5% scavenger loading levels, all plywood met the minimum requirement of 7500 N/mm^2 for exterior applications, reinforcing the suitability of these boards for demanding use cases.

Furthermore, the addition of the scavenger material had a noticeable impact on bond quality, as evidenced by the results of the Glue Shear Strength (GSS) tests. It is important to highlight that the GSS values of the specimens remained consistently above the minimum standards specified. This underscores the effectiveness of the scavenger in enhancing adhesive strength, further enhancing the overall quality and durability of the plywood.

Additionally, the results of cyclic tests carried out as per MR and BWP grade, in accordance with IS: 848(2006) standards, indicated that PF-bonded plywood passed the prescribed standards. This outcome emphasizes the plywood's ability to withstand cyclic loading conditions, reinforcing its suitability for applications where durability and reliability are paramount.

Incorporating the scavenger in the plywood formulation improves its mechanical properties, such as MOR and MOE, enhances adhesive strength, and ensures compliance with rigorous industry standards, making it a promising choice for various applications, including those demanding high-performance characteristics.

Table 5 Mechanical properties of plywood

Sample name	MOR along (N/mm^2)	MOR across (N/mm^2)	MOE along (N/mm^2)	MOE across (N/mm^2)	GSS (N) (Dry)
PF0	54.62	40.81	7555	4508	1602
PF1	54.88	42.62	7600	4581	1722
PF2	55.62	43.68	7625	4602	1748
PF3	57.81	44.15	7686	4686	1786
PF4	58.25	45.08	7691	4711	1806
PF5	58.61	46.52	7710	4722	1825

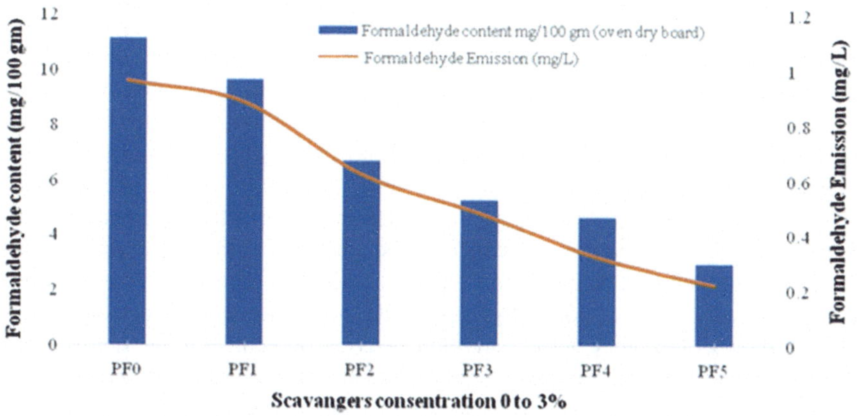

Fig. 2 Formaldehyde emission and content

3.4 Effect of the Scavenger on Formaldehyde Emission

As shown in Fig. 2, the formaldehyde emission data highlights the significant impact of scavenger type and concentration on reducing formaldehyde emissions. The findings reveal that the lowest formaldehyde emission, recorded at 0.22 mg/L, was achieved in the plywood specimen containing 3% (PF5) Polymate-777P. This result was obtained following the testing protocol specified in JIS A 1460. When tested according to IS 13745, the formaldehyde content was found to be 2.96 mg/100 gm board for the same plywood specimen. These results clearly indicate the remarkable efficacy of Polymate-777P in reducing formaldehyde emissions in plywood, making it an effective solution for formaldehyde mitigation in producing plywood.

Moreover, it was observed that the formaldehyde emission values of all plywood boards tested in the study met the requirements of the E_1 grade, as per the Japanese standard JIS A 5908, across all scavenger concentrations. The E_1 grade stipulates a maximum formaldehyde emission of 1.5 mg/L, and the data consistently demonstrated compliance with this stringent standard.

Furthermore, when the plywood boards contained scavenger concentrations of 1.5% (PF3), 2% (PF4), and 3% (PF5), they not only met the requirements of the E_1 grade but also exceeded expectations by achieving compliance with the even stricter E_0 grade. The E_0 grade mandates a maximum formaldehyde emission of 0.5 mg/L, and the data indicate that these scavenger concentrations successfully achieved this exceptional level of formaldehyde reduction.

These results underscore the effectiveness of Polymate-777P as a formaldehyde scavenger in plywood manufactured using phenolic resin and highlight its potential for enhancing the environmental sustainability of plywood production by significantly reducing formaldehyde emissions. This study contributes valuable insights for the industry in its pursuit of safer and more eco-friendly plywood manufacturing processes.

The likely mechanisms underlying the reduction of formaldehyde emission from plywood through the use of a scavenger can be attributed to the scavenger's strong affinity for emitted formaldehyde molecules. This affinity arises from the presence of two adjacent phenolic hydroxyl groups (–OH) on the aromatic ring, which facilitates an accelerated nucleophilic addition reaction. Specifically, the non-bonded electron lone pair of OH groups initiates an attack on the carbonyl groups ($C = O$) in formaldehyde, leading to the formation of a methylol structure ($RO{-}CH_2{-}OH$). Subsequently, the reaction progresses as the other lone pair present on the adjacent ring-OH attacks the methylol group via an intramolecular interaction.

4 Conclusion

The scavenger called Polymate-777P can improve the mechanical properties of plywood and effectively reduce formaldehyde emissions from plywood panels by more than 50%. The study's findings indicate that Polymate-777P can be used at a concentration of 1.0% or more as a scavenger for PF resin to produce plywood panels with lower formaldehyde emissions that meet the strict requirements for outdoor usage. Additionally, when Polymate-777P was employed as a scavenger at concentrations of 1.5% or higher, the manufactured plywood met the Japanese industrial standard F-Three Star emission requirement ($\leq$0.50 mg/L) and European E0 classes. Furthermore, as the scavenger content increased from 0.5% to 3%, the MOR consistently rose, escalating from an initial 54.62 N/mm^2 to a peak of 58.61 N/mm^2. To comply with strict formaldehyde emission limits, it is advised to optimize scavenger concentrations further and incorporate them into wood composite manufacturing processes. These findings contribute to the development of wood composite production methods that are both health- and environmentally-conscious.

Acknowledgements Financial support for this study was provided by a grant from M/S –Patina Chemicals, Kolkata (West Bengal).

Statements and Declarations

Competing Interests and Funding: NA.

Conflicts of interest/Competing interests: The authors declare they have no financial interests.

Ethics approval: Not applicable.

Consent to participate: Not applicable.

Availability of data and materials: Material and all supplementary data are presented in Appendix I including Figs. 3 and 4.

Code availability: Not applicable.

Author contributions All authors contributed to the study's conception and design. Material preparation, data collection, and analysis were performed by S.C. Sahoo and Shibu C. The first draft of the manuscript was written by S.C. Sahoo and Shibu C, and all authors commented on previous versions of the manuscript. All authors read and approved the final manuscript.

Appendix I: Supplementary Data

See (Figs. 3 and 4).

Fig. 3 Material images

patina™
Trust forever... Smile forever...

PATINA CHEMICALS
NON HAZARD CHAMICALS & ALLIED PRODUCTS

POLYMATE -777-P (Formaldehyde scavenger for phenolic resin)

Formaldehyde is a one of the most important starting materials in the chemical industry. It is used for example in the synthesis of urea formaldehyde resins, melamine resins, phenol formaldehyde resins,The issue of formaldehyde emission from composite wood products has been a concern, since the gas was deemed carcinogenic as per IARC. Europe has a set of standards for grading the plywood as per formaldehyde emission level, while India is yet to have any set of standards for plywood. Exporter prefers to to follow the European standards It is unclear how the limits of formaldehyde emissions for wood panels will change in different parts of the world in the future. The wood-based panel industry has so far reacted by applying appropriate formaldehyde scavengers (catchers). As a volatile organic compound (VOC), most of the formaldehyde is normally emitted from panels during production. There are two formaldehyde sources when producing wood panels: formaldehyde that might be contained in the adhesive and that in the wood material itself. The emissions can be reduced by either using formaldehyde scavengers.Formaldehyde scavenger is added directly to resin to reduce formaldehyde emission . Because of the aforementioned risks for human health there was a strong need for effective scavengers for formaldehyde.

Polymate -777-P (scavenger for phenolic resin)

It relates to a new formaldehyde scavengers, for reducing the amount of formaldehyde by adding these scavengers in phenolic resin. Basically it is an additive for plywood adhesive made by with high level formaldehyde catcher specialty chemicals like lignin, caprolactam, PTSA with some organic/inorganic additive having amino group. It is used during synthesis of the resin to active fully without having any adverse reaction with the resin or alter the rheological properties of the resin.

Scavanger Specification :-.
Appearance – Light brown colored fine powder
pH – 6.0 - 6.5 (1% solution)
Ionic Character – Non ionic
Activity -100%
Solubility _100%
Dosage – 0.5 - 1% as per requirement
Shelf life - More than 06 months
Product description
IDENTITY:- Formaldehyde scavenger for wood adhesive,
1.CHEMICAL PRODUCT AND COMPANY
IDENTIFICATION PRODUCT NAME:- POLYMATE -777-P
MANUFACTURER:- PATINA CHEMICALS
2. COMPOSITION/INFORMATION ON INGREDIENTS
CHEMICAL NAME: - Formaldehyde scavenger is an additive composed by some speciality chemicals like Caprolactam, PTSA, Lignin with some inorganic and organic additives having amino as a functional terminating group .
3. HAZARDS IDENTIFICATION
Non hazardeous ,may be harmful if inhaled .
4. FIRST AID MEASURES
GENERAL ADVICE: Wash off immediately with soap and plenty of water. In the case of respirable dust ,use self- contained breathing apparatus and dust impervious protective suit. Use personal protective equipment.
SKIN CONTACT: Remove contaminated clothes and shoes, rinse skin with plenty of water or shower. Use soap to help assure removal.
EYE CONTACT: Remove any contact lenses at once. Flush eyes well with flooding amounts of

VILL-GANGARAMPUR, P.O.-DAKSHIN ALIPUR, P.S.-BISHNUPUR, DIST.-SOUTH 24 PARGANAS, WEST BENGAL-743503

Fig. 4 Polymate-777P MSDS

patina™
Trust forever...
Smile forever...

PATINA CHEMICALS
NON HAZARD CHAMICALS & ALLIED PRODUCTS

running water for at least 15 minutes. Assure adequate flushing by separating the eyelids with sterile fingers. If irritation persists, consult a physician.
INGESTION: Rinse mouth, give plenty of water to dilute the substance. Never give anything by mouth to an unconscious person. Consult a phy
5. HANDLING AND STORAGE
HANDLING: Avoid contact with eyes, skin and clothing. Avoid prolonged or repeated exposure. Handle material with suitable protection.
6.STORAGE:- Store away from sunlight in well- ventilated dry place at room temperature. Keep container tightly closed.
8. PHYSICAL AND CHEMICAL PROPERTIES
Appearance – Light brown colored fine powder
pH – 6.0 - 6.5 (1% solution)
Ionic Character – Non ionic
Activity -100%
Solubility _100%
Dosage – 0.5 - 1% as per requirement
Shelf life - More than 06 months
9 . Transport Information
Proper Shipping Name: None
Non - hazardous for transport: this substance is considered to be non - hazardous for transport.
10. DISPOSAL CONSIDERATION
Dispose in a hazardous- waste site in accordance with all applicable regulations.
11. OTHER INFORMATION
The information above is believed to be accurate and represents the best information currently available to us. However, we make no warranty of merchantability or any other warranty, express or implied, with respect to such information, and we assume no liability resulting from its use. Users should make their own investigations to determine the suitability of the information for their particular purposes.

VILL-GANGARAMPUR, P.O.-DAKSHIN ALIPUR, P.S.-BISHNUPUR, DIST.-SOUTH 24 PARGANAS, WEST BENGAL-743503

Fig. 4 (continued)

References

1. Antov P, Savov V, Krišťák L, Réh R, Mantanis GI (2021) Eco-friendly, high-density fiberboards bonded with urea-formaldehyde and ammonium lignosulfonate. Polymers 13(2):220. https://doi.org/10.3390/polym13020220
2. Ashaari Z, Lee SH, Radzali MNE (2015) Effects of ammonium carbonate post treatment on phenolic resin treated sesenduk (*Endospermum diadenum*) wood. Sains Malays 44(7):987–994. https://doi.org/10.17576/jsm-2015-4407-10
3. Dorieh A, Selakjani PP, Shahavi MH, Pizzi A, Ghafari Movahed S, Farajollah Pour M, Aghaei R (2022). Recent developments in the performance of micro/nanoparticle-modified urea-formaldehyde resins used as wood-based composite binders: a review. Int J Adhes Adhes 114(103106):103106. https://doi.org/10.1016/j.ijadhadh.2022.103106
4. Hussin MH, Abd Latif NH, Hamidon TS, Idris NN, Hashim R, Appaturi JN, Brosse N, Ziegler-Devin I, Chrusiel L, Fatriasari W, Syamani FA, Sedliačik J (2022). Latest advancements in high-performance bio-based wood adhesives: a critical review. J Mater Res Technol 21, 3909–3946. https://doi.org/10.1016/j.jmrt.2022.10.156
5. Kariuki SW, Wachira J, Kawira M, Murithi G (2019) Formaldehyde use and alternative biobased binders for particleboard formulation: a review. J Chem 2019:1–12. https://doi.org/10.1155/2019/5256897
6. Kristak L, Antov P, Bekhta P, Lubis MAR, Iswanto AH, Reh R, Sedliacikh J, Savov V, Taghiyari HR, Papadopoules AN, Pizzi A, Hejna A (2023) Recent progress in ultra-low formaldehyde emitting adhesive systems and formaldehyde scavengers in wood-based panels: a review. Wood Mat Sci Eng 18(2):763–782. https://doi.org/10.1080/17480272.2022.2056080
7. Sahoo SC, Sil A, Riya R, Solankip T (2021) Synthesis and properties of UF/pMDI hybrid resin for better water resistance properties of interior plywood. Int J Innov Sci Eng Technol 8:148–158
8. Sahoo SC (2023) Synthesis and properties study of the urea formaldehyde resin modified with Pmdi used for manufacturing of wood composite products. Eur Chem Bull 12(10):9875–9880
9. Sahoo SC, Amitava S, Dharm P (2020) Effect of nano-additive on the bonding strength and formaldehyde emission of the plywood adhesive during manufacturing of wood based panel products. Int J Innov Technol Explor Eng 10(2):158–161. https://doi.org/10.35940/ijitee.b8306.1210220
10. Sahoo SC, Tudu R, Pandey CN, Khatua PK (2015) Green adhesive-using glyoxalated lignin and hydrogenated resin as ecologically safe wood adhesive for interior plywood. Int J Prev Control Ind Pollut 1(1):11–15
11. Saito K, Hirabayashi Y, Yamanaka S (2021) Reduction of formaldehyde emission from urea-formaldehyde resin with a small quantity of graphene oxide. RSC Adv 11(52):32830–32836. https://doi.org/10.1039/d1ra06717f
12. Selakjani PP, Dorieh A, Pizzi A, Shahavi MH, Hasankhah A, Shekarsaraee S, Ashouri M, Movahed SG, Abatari MN (2021) Reducing free formaldehyde emission, improvement of thickness swelling and increasing storage stability of novel medium density fiberboard by urea-formaldehyde adhesive modified by phenol derivatives. Int J Adhes Adhes 111(102962):102962. https://doi.org/10.1016/j.ijadhadh.2021.102962

Valorization of Waste Office Paper for Extraction of Microcrystalline Cellulose and Its Characterization

Muhil Raj Prabhakar and Paramasivan Balasubramanian

Abstract Waste paper classified as municipal solid waste is often dumped in landfills or incinerated. As a low-cost source of lignocellulosic elements, the waste paper can be converted into cellulose microcrystals. Microcrystalline cellulose (MCC) is becoming increasingly important as an alternative to non-renewable and limited fossil fuels due to its broad use in sectors such as food, pharmaceutical, cosmetic, and polymer composites. Because of its sustainability, non-toxicity, affordability, biodegradability, outstanding mechanical properties, and high surface area, MCC has steadily earned a greater amount of attention in recent years. To meet the growing need for manufacturing new types of MCC-based products on an industrial scale, novel sources, separation procedures, and treatments are continuously being developed. This study aims to extract MCC from waste office paper as a novel source through acid hydrolysis. The synthesized MCCs were subjected to structural and compositional characterization. The MCCs obtained from both sources exhibited a crystalline nature with a particle size distribution of 0.6–1.1 μm, and the length-to-cross-section ratio was found to be 10:1. The FTIR analysis revealed the absence of peaks corresponding to lignin in the extracted MCC. The extracted MCC showed a yield percentage of 79.5 ± 1.2%. Thus, waste office paper can act as a potential and sustainable source for extraction of MCC for application in various fields.

Keywords Microcrystalline cellulose · Waste paper · Valorization · Sustainability · Physio-chemical properties

1 Introduction

The escalating global concerns regarding environmental sustainability and the pressing need for innovative waste management solutions have prompted a heightened focus on the valorization of waste materials. In this context, waste office paper,

M. R. Prabhakar · P. Balasubramanian (✉)
Bioenergy and Environmental Laboratory, Department of Biotechnology and Medical Engineering, National Institute of Technology Rourkela, Rourkela, Odisha, India
e-mail: biobala@nitrkl.ac.in

© The Author(s), under exclusive license to Springer Nature Singapore Pte Ltd. 2024
M. Suar et al. (eds.), *Advancements in Polymer Technology*, Springer Proceedings in Materials 57, https://doi.org/10.1007/978-981-97-6333-7_8

an abundant byproduct of administrative and commercial activities, emerges as a substantial yet underutilized resource that, if harnessed effectively, could contribute significantly to sustainable practices. Paper waste disposed of as municipal solid waste can have the same negative consequences on economic growth and environmental health as other types of municipal solid waste [19]. The pulp and paper industry is the third most significant contributor to soil, air, and water pollution among industrial sectors [10]. Cellulose fiber composition ranges from 90 to 99% of conventional paper-based products. Cellulose fibers serve as both the primary structural element and the component in its entirety that has the most significant influence on the end-use attributes of the product. Paper products have chemical and physical characteristics that are influenced by a network of cellulose fibers that are capable of self-bonding [14]. However, cellulose is a linear polysaccharide composed of repeating β-D-glucose units linked by β-1, 4-glycosidic bonds. The chemical composition is characterized by the glucose monomers, forming a linear chain with a high degree of polymerization. The abundant hydroxyl groups on each glucose unit contribute to cellulose's unique physical and chemical properties [6].

Rising paper consumption contributes to environmental challenges, but also offers opportunities for resource recovery. Waste office paper, comprising 17% of industrial waste globally, can be repurposed for valuable material extraction. Additionally, discarded paper constitutes 26% of municipal solid waste, contributing to environmental pollution from pulp and paper production [8]. In India, 1.2–2.4 million metric tons of newspapers and 2.4–4.3 million metric tons of cardboard and mixed paper are recycled annually, accounting for 30–60% of total production [11]. Cellulose from waste paper can act as a valuable and economical source for producing cellulose-based products. Currently, lignocellulosic biomass is considered a promising source of cellulose, yet consuming lignocellulosic material to produce cellulose can increase the existing deforestation rate globally. Bacterial cellulose is a promising alternative source, however, its production is costly and demands a sterile environment with substantial manpower. Techno-economic analysis indicates that 89% of operating costs primarily stem from facility maintenance and labor [2, 9]. Thus, considering waste paper as a potential source of cellulose can be economical and eco-friendly, reducing the deforestation rate.

Microcrystalline cellulose (MCC), a derivative of cellulose, exhibits unique properties such as high surface area, mechanical strength, and biodegradability, making it an attractive material for applications in pharmaceuticals, food, packaging, and other fields. Recent articles have explored the potential of using waste paper to produce various valuable products other than recycled paper, such as the implementation of superhydrophobic aerogel as a thermal protecting cooler for building areas [20] and an adsorbent aerogel for eliminating organic chemical pollutants and spills of oil from water [12]. In addition, microorganisms were able to transform paper waste into fuels, including bioethanol, biodiesel, and biogas [1]. MCC, a highly pure industrial polymer, is expected to grow at a significant rate of 6.3% from 2023 to 2033, reaching a market value of US $1794.1 million due to rising global demand. Production of MCC from agro-residues involves more stringent chemicals at higher volumes and the degradation of natural resources [17]. Wanrosli et al. [18] utilized oil palm empty fruit bunches as a source of MCC for the synthesis of cellulose phosphate. Bhandari et al. [3] synthesized MCC from carpenter waste through chemical treatment, and the MCC particles were found to possess a crystallinity index in the range of 67.56 to 68.85 with an average particle size ranging from 96.42 to 437.83 nm. The production of MCC from waste paper can be done more cheaply and with less reliance on agricultural waste. Furthermore, the synthesis of MCC from diverse waste paper sources can help to convert municipal solid waste into a value-added product. Even though used office paper can produce MCC, there is a clear research gap regarding the value-adding of this particular waste stream. This study aims to close the gap by thoroughly researching the extraction of MCC from waste office paper and providing a thorough analysis of the obtained material. The motivation behind this study lies in the dual objectives of mitigating environmental impact by repurposing waste and advancing the utilization of MCC in sustainable material development.

To extract MCC, a versatile and biodegradable substance with promising uses across numerous industries, waste office paper is first valued. This research study explores the extraction of MCC from waste papers, which does not require any severe chemical treatment and can be done more cheaply than from any other source, unlike the paper production process, which requires a strict chemical treatment to produce high-quality papers. The surface and chemical characteristics of the manufactured MCCs made from waste office paper are also examined in this work.

2 Materials and Methods

The waste office paper, in general, consists of 85–99% cellulose and 1% lignin with residual ash of about 0–10% [4, 21], making it a suitable source for extraction of MCC. The waste office paper was washed with distilled water to remove any impurities present and ground using a household grinder for 5 min. The ground pulp was washed again with water to remove impurities and then dried overnight at 50 °C. The dried paper pulp was then subjected to the extraction of MCC.

2.1 Synthesis of MCC from the Waste Office Paper

As the paper production itself involves a stringent chemical treatment to achieve good quality papers, the synthesis of MCC from waste papers does not require any harsh chemical treatment and can also be synthesized in a more economical way than any other source. The chemical method of synthesizing MCC was performed with the help of organic acids and alkali. This study followed the protocol reported by Takagi et al. [16] with some modifications. The paper pulp was treated with 2% (w/v) NaOH at 100 °C for 3 h, followed by acid hydrolysis of 30% (v/v) H_2SO_4 for 30 min at 50 °C. The resultant cellulose was washed, neutralized, and subjected to bleaching using NaOH and H_2O_2 for 1 h. The experiment was done in triplicates and the yield was calculated using Eq. (1).

$$Yield(\%) = \frac{W_f}{W_i} \times 100 \quad (1)$$

where W_i is the initial weight of the sample and W_f is the weight of the resultant MCC.

2.2 Structural Characterization of MCC

The surface morphology of the produced MCCs was examined using a FlexSEM 1000 (Hitachi) Scanning Electron Microscope (SEM) and an accelerating voltage of 10 kV, with gold coating before analysis. The Energy Dispersive X-ray Spectroscopy (EDS) was performed to analyze the elemental composition of the MCC. The Bruker AXS D8 Advance with Davinci Design was used to perform X-ray diffraction analysis (XRD) at a scan rate of 10^0 min^{-1} in the 2θ range of 10^0 – 110^0 utilizing a cobalt lamp to determine the crystal properties. The Scherrer equation (Eq. 2) was employed to compute the crystal lattice,

$$L_{h,k,l} = \frac{K\lambda}{\beta cos\theta} \quad (2)$$

where K = 0.94 (Scherer's constant), λ = 0.15406 nm (wavelength of incident radiation), β is the full width at half maximum of the peaks (rads), and θ is the diffraction angle of obtained peaks (rads) [13]. The crystallinity index is the ratio of the area under crystalline peaks in the diffractogram to the total area under the diffractogram. The percentage crystallinity index (% CrI) was also calculated using Segal's equation (Eq. 3),

$$\%CrI = \frac{I_{200} - I_{AM}}{I_{200}} * 100 \quad (3)$$

where $A_{crystal}$ is the area of the crystalline peaks in the X-ray diffractograms, and A_{total} is the total area of the X-ray diffractograms. The particle size distribution of synthesized MCCs was calculated through the dynamic light scattering method using Zetasizer Ver. 7.13 (Malvern Panalytical).

2.3 Chemical Characterization

The Fourier transform infrared spectrophotometer (FTIR) was utilized to analyze the chemical composition of the synthesized MCCs using Bruker ATR–FTIR spectrometer. The spectral transmittance of the materials had been assessed within the range of wavenumbers encompassing from 4000 cm^{-1} to 500 cm^{-1}.

3 Results and Discussion

The utilization of paper waste as a source of cellulose presents a promising avenue, as it offers a rich source of cellulose that can be extracted without the need for intensive chemical treatment, unlike the extraction process from plant sources. The extraction of MCC was carried out from waste office paper through acid hydrolysis, resulting in a pale grey-colored powder, as shown in Fig. 1.

Since there was no significant interference from minor quantities of lignin, O-MCC's synthesis was carried out using a simplified methodology. The SEM analysis was performed to analyze the surface morphology of the extracted MCC and the elemental composition of the extracted MCC was estimated using EDS (Fig. 2). Some of the extracted MCC structures were found to be self-agglomerated with individual polymer particles, clearly seen in Fig. 2a. The particles that didn't agglomerate were found to be rod-shaped without any fibrous regions at the end (Fig. 2b), providing additional proof that the extracted MCC is crystalline. The obtained individual MCC crystals possessed a length of around 70–100 μm with a cross-section of around 7–10 μm measured through ImageJ software [15] from the SEM images, and the length to-cross-section ratio was found to be 10:1. The EDS was performed to analyze the elemental composition of the MCC (Fig. 2c), and it was observed that a trace amount of sodium and chloride had been seen, which may be due to the bleaching

Fig. 1 The image of synthesized MCC from office paper

step followed by hydrolysis (Table 1) in addition to those traces of sulfur is also observed. The extraction process of MCCs was repeated thrice, and the percentage yield of the obtained MCC was calculated as 79.5 ± 1.2%, respectively.

Cellulose consists of several hydroxyl (-OH) groups that contribute to the formation of both intra and inter-molecular hydrogen bonds. Because of the hydrogen bonds, molecules are arranged in an organized manner, and cellulose has a crystalline structure. Figure 3 displays the diffraction pattern for the MCC that was synthesized compared with the diffraction pattern of raw waste paper. The waste office paper was observed to be highly amorphous. X-ray diffractogram of MCC exhibited 2θ values around 16–19°, 26°, and 40°, ascribed to (1 0 1), (2 0 0), and (0 4 0) crystallographic planes, respectively was calculated through Eq. (2) with slight variations in the intensities due to different levels of crystallinity. The crystallinity index was

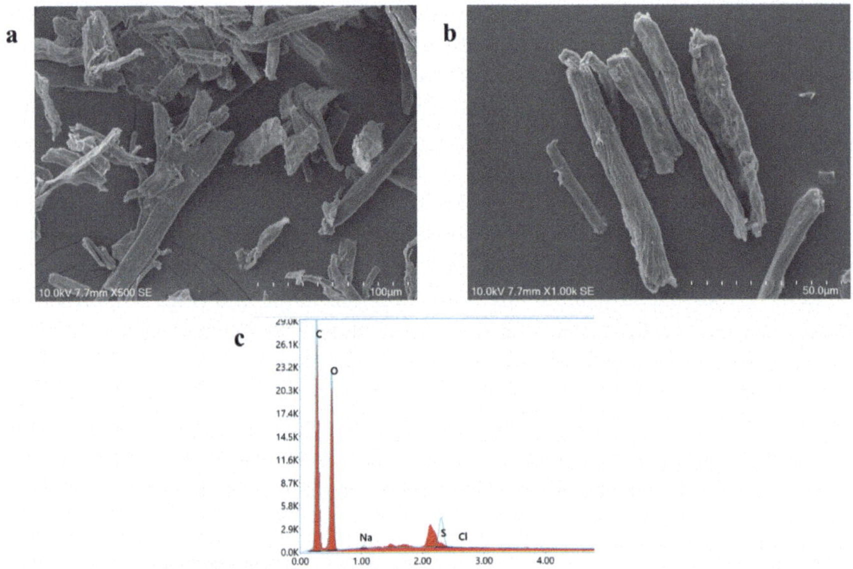

Fig. 2 The SEM images of MCC at **a** 500 × and **b** 1000 × resolution and elemental analysis using EDS (**c**)

Table 1 Elemental composition of extracted MCC through EDS

Element	Weight %	Atomic %
C	63.2	70.3
O	34.2	28.6
Na	0.2	0.1
S	2.4	1.0
Cl	0.0	0.05

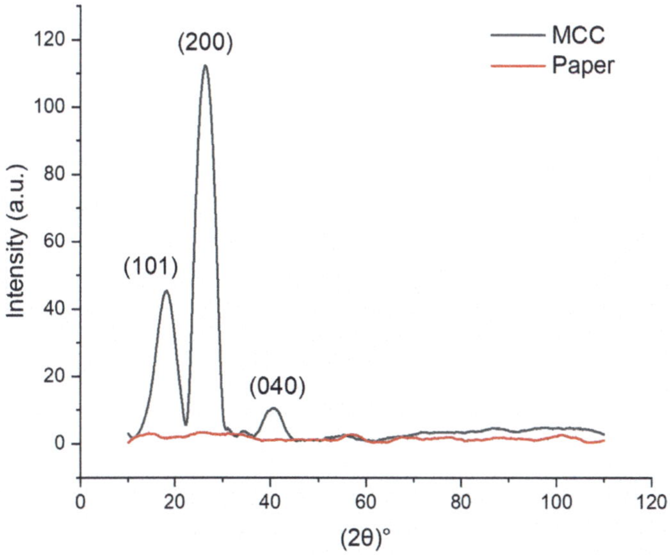

Fig. 3 X-ray diffractograms of MCC obtained through chemical treatment

calculated using Segal's formula Eq. (3) and found to be 73.6%. [7] extracted crystalline cellulose from a waste newspaper which exhibited a similar diffraction pattern with a crystallinity index of about 70.21%. However, the yield of cellulose from the waste newspaper was very low around 35% as the treatment involved more stringent conditions (61% H_2SO_4) when compared to the present study.

DLS is a successful technique for approximately determining the particle size distribution for substances ranging from micro to nano-sized. In this analysis, MCCs are represented as spheres that are moving around using Brownian motion. determining the hydrodynamic radius (RH) of the particles by measuring their translational diffusion coefficient, which is then translated to RH. The synthesized nanoparticles were exposed to ultrasonication to disperse them in an aqueous phase, and then the particle size distribution was calculated through the dynamic light scattering method. The particle-size distribution of extracted MCC is depicted in Fig. 4. It was observed that the particle size distribution ranged from 0.6 to 1.1 μm.

The functional groups of extracted MCC were identified employing FTIR analysis and compared with the FTIR of office paper (Fig. 5). The FTIR spectrum shows that they all have absorption peaks at 3290 cm^{-1}, 2880 cm^{-1}, 1690 cm^{-1}, 1546 cm^{-1}, 1314 cm^{-1}, and 1026 cm^{-1} in common, which represent the –OH stretching vibration peak, the –CH stretching vibration peak, the -C–O–C stretching vibration characteristic peak, and alienation –bonds, respectively. The position of the basic characteristic peak of cellulose has not changed, indicating that the molecular structure of the MCC remained mostly constant during the hydrolysis process, and the quality is essentially consistent. The absorbance peak at 1026 cm^{-1} was found to be sharper due to regeneration, indicating that the C-O in the amorphous area becomes stronger, and the

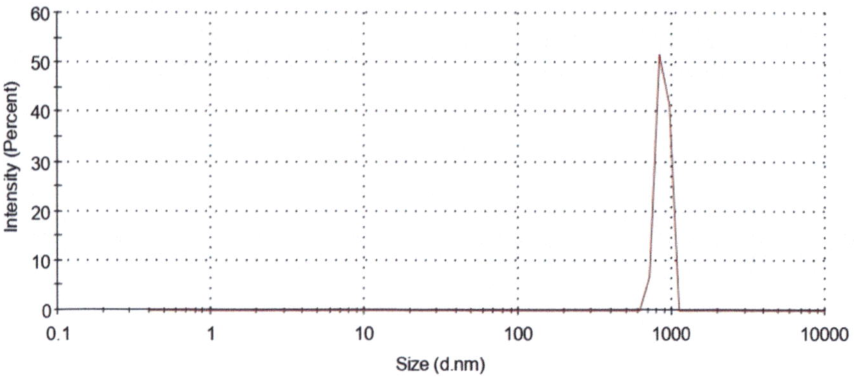

Fig. 4 Particle size distribution analysis of MCC through dynamic light scattering method

cellulose crystal shape changes from type I to type II [5]. In addition to that, the characteristic peaks for lignin, usually found between 1163 and 1275 cm^{-1} corresponding to the C-O stretching of esters, vinyl ether linkages, and alkyl aryl ether were found to be missing, confirming the absence of lignin in the extracted MCC [3]. Since lignin is found to be in trace amount in waste office paper, its corresponding peaks between 1595 cm^{-1}, 1505 cm^{-1}, and 1247 cm^{-1} have been masked by the other peaks, thus the derivative plot has exhibited its presence with very weak peaks.

MCCs are commonly utilized in pharmaceutical tablet formulations and as emulsifiers in the food industry due to their characteristic of being both odorless and tasteless. The purity of MCCs derived from lignocellulosic biomass is of a high

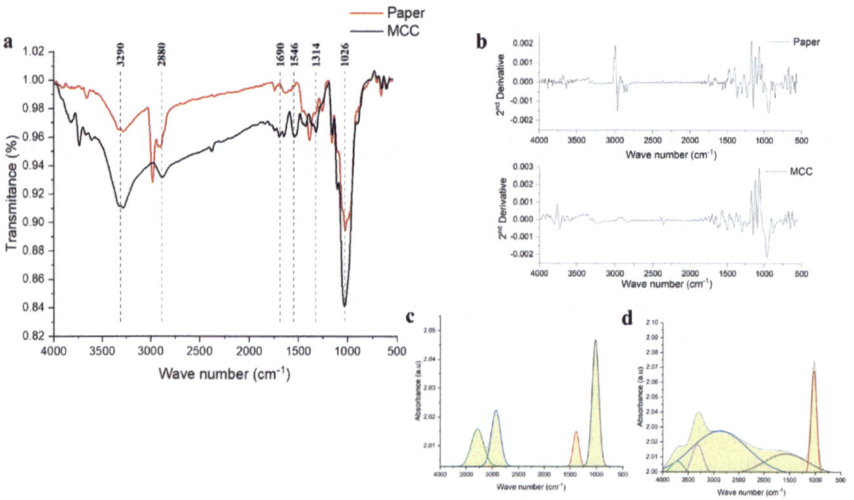

Fig. 5 FTIR spectrum of MCC (**a**), derivative plot (**b**), and deconvolution plot (**c**) & (**d**) of raw waste office paper and extracted MCC through chemical treatment

degree, thereby meeting the requisite standards for pharmaceutical and food applications. However, the MCCs that were synthesized in this study were sourced from waste materials that do not meet the aforementioned rigorous standards. Consequently, it has the potential to serve as a thickening and emulsifying agent in the paint industry while also acting as a filler substance in the paper manufacturing sector. The synthesized substance showcases its potential as a substitute for lignocellulosic-based MCCs of high purity in the manufacturing of thermostable insulators, carbon scrubbers, and other related products.

4 Conclusion

The current study investigated the possibility of using waste office paper as a source for the extraction of MCC. The extracted MCC from waste office paper showed identical structural and compositional properties similar to that of MCCs synthesized from agro-residues. The absence of lignin peaks in FTIR The application of an efficient chemical treatment to waste newspaper pulp, as demonstrated by FTIR analysis of the MCC, resulted in the total elimination of hemicellulose and lignin from the material. The percentage crystallinity index of MCC was approximately 73.6%. The extracted MCC possessed an average particle size of 0.6–1.1 μm. Therefore, waste office paper could act as a potential source for the extraction of MCC, achieving more sustainability in comparison with other lignocellulosic sources in various material processing industries.

Declarations

Ethical approval

This declaration is not applicable.

Data availability

All data supporting the findings of this study are available within the paper.

Funding Sources

No external funding.

Competing interest

The authors declare that they have no known competing financial interests or personal relationships that could have appeared to influence the work reported in this paper.

Author contributions

All authors contributed to the study's conception and design. Material preparation, data collection, and analysis were performed by Muhil Raj Prabhakar. The preparation of the manuscript and its proofreading were done by Muhil Raj Prabhakar and Balasubramanian Paramasivan. All authors read and approved the final manuscript.

Acknowledgements The authors are grateful to the Ministry of Education, Govt of India, for full-time PhD fellowship assistance and the Department of Biotechnology and Medical Engineering, NIT Rourkela, for providing their research facilities.

References

1. Al-Azkawi A, Elliston A, Al-Bahry S, Sivakumar N (2019) Waste paper to bioethanol: current and future prospective. Biofuels Bioprod Biorefin 13(4):1106–1118. https://doi.org/10.1002/BBB.1983
2. Behera B, Laavanya D, Balasubramanian P (2022) Techno-economic feasibility assessment of bacterial cellulose biofilm production during the Kombucha fermentation process. Biores Technol 346:126659. https://doi.org/10.1016/J.BIORTECH.2021.126659
3. Bhandari K, Adaval A, Bhattacharyya AR, Maulik SR (2022) Synthesis of microcrystalline cellulose from carpenter waste and its characterizations. J Nat Fibers 1–15. https://doi.org/10.1080/15440478.2020.1788688
4. Byadgi SA, Kalburgi PB (2016) Production of bioethanol from waste newspaper. Procedia Environ Sci 35:555–562. https://doi.org/10.1016/J.PROENV.2016.07.040
5. Do MH, Van ThiKhuat K, Huynh PT, Nguyen LNT, Do BH, Pham PD, Nguyen HD, Nguyen HM, Thach UD (2022) Synthesis, characterization, and antibacterial activity of amino-functionalized microcrystalline cellulose derivatives from cotton fibers. Biomass Convers Biorefinery 1:1–9. https://doi.org/10.1007/S13399-022-02391-7/FIGURES/6
6. Gelaw BB, Kasaew E, Belayneh A, Tesfaw D, Tesfaye T (2023) Review of the sources, synthesis, and applications of nanocellulose materials. Polym Bull 2023:1–23. https://doi.org/10.1007/S00289-023-05061-4
7. Guan Y, Li W, Gao H, Zhang L, Zhou L, Peng F (2021) Preparation of cellulose nanocrystals from deinked waste newspaper and their usage for papermaking. Carbohydr Polym Technol Appl 2:100107. https://doi.org/10.1016/J.CARPTA.2021.100107
8. Haile A, Gelebo GG, Tesfaye T, Mengie W, Mebrate MA, Abuhay A, Limeneh DY (2021) Pulp and paper mill wastes: utilizations and prospects for high value-added biomaterials. Bioresour Bioprocess 8(1):1–22. https://doi.org/10.1186/S40643-021-00385-3/FIGURES/14
9. Laavanya D, Shirkole S, Balasubramanian P (2021) Current challenges, applications, and future perspectives of SCOBY cellulose of Kombucha fermentation. J Clean Prod 295:126454. https://doi.org/10.1016/J.JCLEPRO.2021.126454
10. Mandeep GGK, Liu H, Shukla P (2019) Pulp and paper industry–based pollutants, their health hazards and environmental risks. Curr Opin Environ Sci Health 12:48–56. https://doi.org/10.1016/J.COESH.2019.09.010
11. Nandy B, Sharma G, Garg S, Kumari S, George T, Sunanda Y, Sinha B (2015) Recovery of consumer waste in India–a mass flow analysis for paper, plastic and glass and the contribution of households and the informal sector. Resour Conserv Recycl 101:167–181. https://doi.org/10.1016/J.RESCONREC.2015.05.012
12. Nguyen ST, Feng J, Le NT, Le ATT, Hoang N, Tan VBC, Duong HM (2013) Cellulose aerogel from paper waste for crude oil spill cleaning. Ind Eng Chem Res 52(51):18386–18391. https://doi.org/10.1021/IE4032567/ASSET/IMAGES/LARGE/IE-2013-032567_0008.JPEG
13. Revol JF, Dietrich A, Goring DAI (2011) Effect of mercerization on the crystallite size and crystallinity index in cellulose from different sources. 65(8):1724–1725. https://doi.org/10.1139/V87-288
14. Sahin HT, Arslan MB (2008) A study on physical and chemical properties of cellulose paper immersed in various solvent mixtures. Int J Mol Sci 9(1):78–88. https://doi.org/10.3390/IJMS9010078
15. Schneider CA, Rasband WS, Eliceiri KW (2012) NIH Image to ImageJ: 25 years of image analysis. Nat Methods 9(7):671–675. https://doi.org/10.1038/nmeth.2089
16. Takagi H, Nakagaito AN, Bistamam MSA (2013) Extraction of cellulose nanofiber from waste papers and application to reinforcement in biodegradable composites. J Reinf Plast Compos 32(20):1542–1546. https://doi.org/10.1177/0731684413494109
17. Ventura-Cruz S, Tecante A (2021) Nanocellulose and microcrystalline cellulose from agricultural waste: review on isolation and application as reinforcement in polymeric matrices. Food Hydrocolloids 118:106771. https://doi.org/10.1016/J.FOODHYD.2021.106771

18. Wanrosli WD, Rohaizu R, Ghazali A (2011) Synthesis and characterization of cellulose phosphate from oil palm empty fruit bunches microcrystalline cellulose. Carbohyd Polym 84(1):262–267. https://doi.org/10.1016/J.CARBPOL.2010.11.032
19. Wichai-utcha N, Chavalparit O (2019) 3Rs policy and plastic waste management in Thailand. J Mater Cycles Waste Manag 21(1):10–22. https://doi.org/10.1007/S10163-018-0781-Y/FIGURES/8
20. Yue X, Wu H, Zhang T, Yang D, Qiu F (2022) Superhydrophobic waste paper-based aerogel as a thermal insulating cooler for building. Energy 245:123287. https://doi.org/10.1016/J.ENERGY.2022.123287
21. Zhang Z, Macquarrie DJ, De Bruyn M, Budarin VL, Hunt AJ, Gronnow MJ, Fan J, Shuttleworth PS, Clark JH, Matharu AS (2015) Low-temperature microwave-assisted pyrolysis of waste office paper and the application of bio-oil as an Al adhesive. Green Chem 17(1):260–270. https://doi.org/10.1039/C4GC00768A

SPRINGER NATURE

GPSR Compliance

The European Union's (EU) General Product Safety Regulation (GPSR) is a set of rules that requires consumer products to be safe and our obligations to ensure this.

If you have any concerns about our products, you can contact us on ProductSafety@springernature.com

In case Publisher is established outside the EU, the EU authorized representative is:

Springer Nature Customer Service Center GmbH
Europaplatz 3
69115 Heidelberg, Germany

The manufacturer's authorised representative in the EU is Springer
Nature Customer Service Centre GmbH, Europaplatz 3, 69115 Heidelberg,
Germany. If you have any concerns regarding our products, please
contact ProductSafety@springernature.com

Printed and bound by CPI Group (UK) Ltd, Croydon, CR0 4YY
01/10/2026
02225977-0008